DEPARTMENT OF THE ENVIRONMENT

CENTRAL UNIT ON ENVIRONMENTAL POLLUTION

# LEAD POLLUTION IN BIRMINGHAM

*A Report of the Joint Working Party
on Lead Pollution around Gravelly Hill*

Pollution Paper No. 14

*LONDON*

HER MAJESTY'S STATIONERY OFFICE

© *Crown copyright 1978*
First published 1978
ISBN 0 11 751325 3

This is one of a series of papers to be published for the Department of the Environment on various aspects of pollution. The previous reports were:

1. The Monitoring of the Environment in the United Kingdom (HMSO, 1974).

2. Lead in the Environment and its Significance to Man (HMSO, 1974).

3. The Non-Agricultural Uses of Pesticides in Great Britain (HMSO, 1974).

4. Controlling Pollution (HMSO, 1975).

5. Chlorofluorocarbons and their Effect on Stratospheric Ozone (HMSO, 1976).

6. The Separation of Oil from Water for North Sea Oil Operations (HMSO, 1976).

7. Effects of Airborne Sulphur Compounds on Forests and Freshwaters (HMSO, 1976).

8. Accidental Oil Pollution of the Sea (HMSO, 1976).

9. Pollution Control in Great Britain: How it Works (HMSO, 1976).

10. Environmental Mercury and Man (HMSO, 1976).

11. Environmental Standards. A description of United Kingdom Practice (HMSO, 1977).

12. Lead in Drinking Water. A Survey in Great Britain, 1975–1976 (HMSO, 1977).

13. Tripartite Agreement on Stratospheric Monitoring between France, the United Kingdom and the United States of America (HMSO, 1977).

# FOREWORD

When I set up the Joint Working Party on Lead Pollution around Gravelly Hill, I was conscious not only of the increasing concern, both locally and nationally, about lead pollution, but also that collaboration between professional people with a wide range of responsibilities was required to make rapid progress on a highly emotive topic. I believe that this report amply justifies my view.

The Working Party has concluded that airborne lead levels around the motorway interchange are not exceptional by urban standards, despite the formidable amount of traffic. Similarly, local people have blood lead levels which are not unusual for city dwellers. The studies have not shown any cause for special concern about lead pollution from such concentrations of traffic, though we will continue to keep the whole subject under review.

With my endorsement, the Working Party have also looked more widely at the exposure to lead particularly for children, right across the City. The results in respect of the older children (those of school age) have been reassuring, but there is some indication that there may be a problem of lead intake for some pre-school children living in central areas. Respiration of airborne lead seems not to be responsible for the enhanced exposure so far discovered, but further work is required to establish whether there is a general problem on any scale and what sources of lead are to blame.

Recommendations for such work have been put forward and I endorse these. I also accept the view that a small Steering Committee is required to manage the work and will be discussing appropriate arrangements with all concerned. This particular Working Party has, however, now completed its task.

The work reported here is an important contribution to our knowledge about lead pollution, both in showing that some earlier fears are unjustified, and in having pin-pointed an area where further work is necessary. It is a valuable contribution to policy-making, both in this country and more widely in the European Community. I should like to take this opportunity of thanking all members of the Working Party for their work.

*Denis Howell*

DENIS HOWELL
Minister of State
Department of the Environment

# CONTENTS

# CHAPTER 1

## INTRODUCTION

1. The purpose of this report is to provide a comprehensive record of the work on lead pollution (and its implications for public health) which has been considered by and carried out on behalf of the Joint Working Party on Lead Pollution around Gravelly Hill.* The report therefore deals with a number of aspects of concern associated primarily with airborne and deposited lead, and predominantly with that from petrol-engined vehicles.

2. The presence of lead in our environment and its significance for human health, particularly that of children, has been a contentious topic and one which has caused some public concern for many years. Several countries including the UK (Department of the Environment, 1974), France (La Documentation Francais, 1974), Canada (National Research Council Canada, 1973), and the United States of America (US National Academy of Science, 1972) have carried out reviews of the problem as it exists in those countries, and a number of international organisations including for example the World Health Organisation, the Food and Agriculture Organisation and the Commission of the European Communities have active interests in the subject (see for example, World Health Organisation, 1977). There is, in addition, a vast and varied literature.

3. In December 1971, the then Chief Medical Officer of the Department of Health and Social Security wrote to all Medical Officers of Health drawing their attention to possible environmental hazards associated with lead. The then MOH for Birmingham referred in his reply to lead in air from traffic and described an investigation of blood lead concentrations (an indicator of recent exposure to lead) and air lead concentrations which was just then starting around the Gravelly Hill M6–A38(M) motorway interchange, the area known colloquially as 'Spaghetti Junction'. The reasons which prompted such a survey are perhaps most graphically illustrated by Figure 1, an aerial photograph of the area, and by the realisation that lead emissions from petrol engines are responsible for much the greater part of airborne lead. He also referred to a survey of lead in dust just beginning in the City.

4. The Gravelly Hill project was to be jointly supervised by the University of Aston in Birmingham and City of Birmingham Health Committee, which was also contributing towards the costs. This study was felt to be particularly

---

*The formation of the Joint Working Party is outlined at paragraph 6; the complement is detailed at Appendix A.

important because it was the first in the UK to investigate the effect of the introduction of a major motorway interchange into a city on atmospheric lead concentrations and to attempt to relate the latter to effects on blood lead concentrations in people living nearby.

5. Reports of the surveys were submitted regularly to the Birmingham Health Committee. In early 1974 the results attracted considerable local and national publicity and newspapers carried articles reporting that blood lead concentrations in people living near the interchange had doubled since its opening in May 1972. These reports inevitably caused a great deal of public concern.

6. The Minister of State at the Department of the Environment, the Rt. Hon. Denis Howell, MP, visited Birmingham in early March 1974 and discussed the situation with local officials, members of Government Departments and University staffs. During this visit the Minister initiated the formation of a Joint Working Party on Lead Pollution around Gravelly Hill, on which were to be represented local officials with responsibilities for public health and the environment, members of the Departments of the Environment and Health and Social Security and University staffs. The membership of the Joint Working Party is given at Appendix A; its terms of reference were "to report to the Minister as soon as possible on the situation with regard to:
  i. lead pollution in the environment around the Gravelly Hill interchange; and
  ii. its implications for the health of local people".

7. In its early meetings, a number of papers were considered by the Working Party, which concluded in its First Report* that the available results of the initial blood lead studies which had engendered concern were difficult to interpret because of changes which had been made during the course of the work in sampling and analytical techniques, and in the laboratory used, and because no matched control groups had been studied. Nevertheless, from the small number of samples taken using a fixed sampling and analytical technique, and analysed in the same laboratory, it was possible to conclude that over the period from October 1972/March 1973 to October 1973/January 1974 a small increase (ca 4–5 $\mu$g/100 ml)† had occurred in the mean blood lead concentrations of a group of adult men and women. It was not possible to determine whether this was a consequence of seasonal or climatic factors, nor whether sources of lead other than the interchange might be responsible, but it was emphasised that the latest blood lead concentrations were within the typical physiological range for urban dwellers.

---

*Joint Working Party on Lead Pollution around Gravelly Hill. First Report, Department of the Environment, December 1974.
†Blood lead concentrations are quoted throughout in conventional rather than SI units; the relationship between the two is shown in Appendix B.

8. Because this conclusion is markedly at variance with reports appearing in newspapers at the time (to the effect that blood lead concentrations had doubled) a full consideration of the results available and of the confounding factors in their interpretation is given at Appendix C.

9. The Working Party also considered the work which was continuing on monitoring airborne lead at a number of sites in the vicinity of the interchange. It was concluded in the First Report, in which a summary of the available results was given, that concentrations of airborne lead were in no way exceptionally high by comparison with other urban areas. Further details of this monitoring exercise are given in Chapter 2 of this report.

10. The Working Party also recognised the importance of having adequate data on lead concentrations in dust and it was decided that the survey of dust then in progress under the control of the City Environmental Department should be expanded to include areas around the motorway interchange.

11. Having reviewed the situation in the light of the results available in March 1974, the Working Party made a number of recommendations designed to improve the structure and planning of the project and to extend its basis. A programme of monitoring blood lead concentrations in school children aged 8–14 years, selected from schools across the City, together with a similar study in adults, was organised. Analysis of the samples was carried out by the City Scientific Officer with a number of duplicate samples being analysed by the Institute of Child Health in London. This work was jointly funded by the City, the Department of the Environment and the Department of Health and Social Security. At the same time, monitoring of airborne lead was continued, both around the motorway interchange and, under the City's direction, at sites across the City. Concentrations of lead in dust were also monitored by the then Environmental Department.*

12. A short report of this phase of the work was given in the Interim Report of the Joint Working Party.† The results of the blood lead surveys were considered to be reassuring, since the mean blood lead concentrations in people living near the interchange appeared to have levelled off at values within the typical range for urban dwellers; moreover, the mean concentrations in school children were in the range 12–16 $\mu$g/100 ml, with no child exceeding 35 $\mu$g/100 ml, the upper limit of normality as indicated in the EEC Directive on Biological Screening of the Population for Lead.‡ Air lead concentrations had shown no statistically significant change over the reporting period but concentrations inside houses were found to be substantially the same as those

---

*Now the Environmental Health Department.
†Joint Working Party on Lead Pollution around Gravelly Hill. Interim Report, Department of the Environment, December 1975.
‡Council Directive on Biological Screening of the Population for Lead, Official Journal of the European Communities, Volume 20, No. L105, 28 April 1977, pp. 10–17.

immediately outside. The work on dust had indicated the potential danger to young children of lead contaminated dust in the home. Full details of this programme of work are given in Chapter 2 for the airborne lead and lead in dust, and in Chapters 3 and 4 for the study of blood lead concentrations.

13.  Notwithstanding the reassuring nature of the results obtained on school children, the Working Party recognised the need for an additional study of very young children who might be especially exposed through the ingestion of lead-contaminated dust from sticky fingers and playthings. Accordingly, a study of children under five years of age living throughout the City was organised under the joint sponsorship of the City and the Departments of the Environment and Health and Social Security, and carried out in March 1977. Follow-up work has continued and full details of this work are now given in Chapter 5 and Appendix I of this report.

# CHAPTER 2

## STUDIES OF THE PHYSICAL ENVIRONMENT

**Lead in air***

14. A large number of measurements (5077) of atmospheric lead concentrations have been made at a variety of sites in the City.† The first samples were taken covering the period October 1971–August 1972 from four sites (see Figure 2):

    A Salford Park, roadside site on the heavily used A38. This site was used until May 1972.

    B Slade Road School, some 506 m from Salford Circus. This site was used from January 1972 onwards.

    C Eliot Street School, 580 m from Salford Circus, used from March 1972 to October 1972.

    D Salford Circus, a site within the interchange, 122 m from Salford Park.

Later on, additional sites were set up on gantries spanning the Aston Expressway (E and F in Figure 2) and the A38 at Gravelly Hill (G), and on the roof of some high-rise flats at Murdoch Point on the southern side of Salford Park (H).

15. The samples were collected by drawing air at an average flow rate of 20 litre/min through a bank of sixteen 0·8 μm pore diameter Millipore filters. Twelve hour sampling periods were used, from 0730 to 1930 and from 1930 to 0730, corresponding roughly to 'day-time' and 'night-time' respectively. After exposure the filters were wet-digested with nitric acid/perchloric acid and the lead content of the sample determined using atomic absorption spectrophotometry.

16. Monthly average lead concentrations obtained up to May 1973 are shown in Tables 1, 2 and 3. Whilst comparisons are difficult, because of seasonal variations, it is obvious that airborne lead concentrations are highest at the sites closest to the interchange (Salford Park and Salford Circus). Moreover, it is at these sites that the maximum 12-hour concentrations are found (Table 4) and at which the highest concentrations of lead in the total particulate matter occurred during the early months of the survey (Table 5).

17. Another feature, well illustrated in Figure 3, was the drop in atmospheric lead concentrations at Salford Park (i.e. on the A38) between the autumn of

---

*Work carried out by the Department of Chemistry, University of Aston, Birmingham.
†Some of the results from these studies have been reported elsewhere, Butler and MacMurdo, 1974; Butler, MacMurdo and Middleton, 1974.

1971 and 1972. Before the junction opened, an average of 2,350 vehicles per hour entered Salford Circus on weekdays. This decreased to 1,417 vehicles per hour during May when some traffic began to use the Aston Expressway. Certainly during May there was less traffic passing the Salford Park (A38) monitoring site and, simultaneously, the atmospheric lead concentration fell to less than 1·0 $\mu$g/m³ for the first time since monitoring started. After the junction became operational the number of vehicles entering the complex increased by about 2½ times, to 5,600 per hour, and this is reflected in the data for Salford Circus (Figure 3). No such effect was apparent at the two schools. Weekly total traffic movements, up to the middle of 1974, are shown in Figure 4.

18.   The atmospheric lead concentrations were found to be subject to wide monthly fluctuations which did not correspond closely with the traffic flow and it seemed evident that other factors were influencing the results. During the early studies there was a suggestion that wind speed might be related inversely to the airborne lead concentration and this impression was confirmed by subsequent analysis of the data from Salford Circus. From this analysis it became evident that both wind speed and temperature were important and a predictive equation was developed which had the form:

$$Pb = 2·19 — 0·20WSP — 0·072TMP + 1·1 \times 10^{-6}T$$

where Pb is the air lead concentration in $\mu$g/m³, WSP is the monthly average wind speed in m per sec, TMP is the monthly average temperature in °C, and T is the monthly average number of vehicles. The significance of this equation is discussed at paragraph 21.

19.   Monitoring has continued predominantly at the Salford Circus site, and the monthly average concentrations are shown in Table 6. This site is very near to the carriageways of the interchange, (within about 45 metres) and the airborne lead concentrations are characteristically higher during the day, when the traffic flow is heavy, than at night (Tables 1 and 2). On average, night-time concentrations are about 60 per cent of the day-time values. However, even though the site is so near the source, it is apparent from Table 6 that despite an increase in the traffic flow from about 360,000 vehicles per week before the interchange opened to nearly a million vehicles per week at the present there has been no significant trend in airborne lead concentrations. The explanation presumably lies in the fact that such a diffuse source, occupying a well-ventilated area, is conducive to rapid dispersal and dilution of traffic fumes.

20.   Table 6 also shows that there is a seasonal pattern in air lead concentrations, the highest values being seen in October, November and December, when the concentration on a monthly average basis reaches about 2 $\mu$g/m³. It may be noted that the Commission of the European Communities has made a proposal for air quality standards for lead and this suggests that a value of 2 $\mu$g/m³ should not be exceeded on an *annual* mean basis. Whilst the validity

of this proposal is in doubt, it is clear from Table 6 that the proposed standard would be met even at a site within the interchange itself. Tables 1–3 indicate that at residential and school sites, concentrations are well below the proposed limit.

21. Figure 5 compares the observed data with that predicted from the equation given at paragraph 18 above. For the most part there is good agreement between the two values. However, in exceptionally stable atmospheric conditions, e.g. during temperature inversions, the model breaks down and examples of this are seen in February–March 1973 and November–December 1975. It may readily be shown that the highest lead concentrations will be seen in conditions of atmospheric stability and low wind speed; making reasonable assumptions about peak traffic flow, traffic speed and the lead content of petrol, it may be shown by the methods described by Pasquill (1974) that airborne lead concentrations of about 2·7 $\mu g/m^3$ would be observed at residential sites 300 metres downwind of the interchange. An examination of meteorological data and the relationship between this and other relevant factors such as traffic flow, suggest that high concentrations should only be seen infrequently; this is consistent with Figure 5.

**Comparison of urban and suburban airborne lead concentrations***
22. To compare urban and suburban airborne lead concentrations, pairs of sites were selected. The urban sites (1 and 2) were approximately equidistant from the A34 (Figure 6), in Longlands House and Arden School. Both sites were near a sampling point which the Warren Spring Laboratory had set up alongside the A34 to measure kerbside pollution from motor vehicles (WS in Figure 6). The suburban sites (3 and 4) were in a residential area in postal district 32 (Figure 7). Site 3 was in the Woodgate Valley Association Hall and site 4 in the Milebrook Road Community Centre. Air samples were collected as before from 6 November 1975–11 December 1975 (sites 1 and 2) and from 18 February 1976–16 March 1976 (sites 3 and 4). The results of the analyses are shown in Table 7 from which it will be seen that the suburban mean values are approximately half the urban means, although greater than values obtained in a rural area near Newtown in Montgomeryshire. The day-time mean (over 4 days) in this rural site was 0·16 $\mu g/m^3$ and the night-time mean was 0·20 $\mu g/m^3$. The urban means are similar to those obtained at the school sites around the motorway interchange (see Tables 1, 2 and 3), for comparable times of the year.

**Lead concentrations inside houses***
23. It is important to know how airborne lead concentrations inside houses compare with those outside, since the atmosphere within the house is the one to which people are exposed for most of their day. Three houses have been studied to determine the relationship between internal and external airborne

---

*Work carried out by the Department of Chemistry, University of Aston, Birmingham.

lead concentrations. The first house investigated was situated approximately 160 m from the Salford Circus motorway exit, the second and third were close to the M6 in Perry Barr.

24. Sampling apparatus was installed inside the houses and immediately outside and designed to run simultaneously with that which had been already set up at Salford Circus. From the results (Table 8) it may be seen that lead concentrations inside the house fluctuate similarly to those outside, and indeed in the first house sampled the two concentrations are virtually identical. The fluctuations noted inside and outside the houses closely followed those observed at Salford Circus although both were considerably lower than the Salford Circus values. This is, of course, to be expected since it has been well established that atmospheric lead concentrations decrease rapidly with increasing distance from a roadway (Hartl and Resch, 1972). On the limited evidence available, it seems that houses with double glazing* (house 3, Table 8) are likely to afford little protection from lead in the atmosphere.

25. Blood lead concentrations of two inhabitants from the first house were each 24 $\mu$g/100 ml and those from a single occupant of each of the second and third houses were 19 and 15 $\mu$g/100 ml respectively, values well in keeping with those from other adults in the City (see Chapter 4).

**The influence of noise barriers on the dispersion of lead†**
26. Noise barriers are placed alongside some motorways to reduce noise levels when these affect people living nearby. They are continuous structures which deflect sound waves away from nearby buildings. The Working Party considered whether these barriers might influence the concentrations of airborne lead at houses close to motorways. It was not necessary to carry out research in the Birmingham area, since the work of the Transport and Road Research Laboratory on the M4 motorway at Heston was made available to the Working Party.

27. Measurements of airborne lead concentrations were made outside houses adjacent to the M4 both before and after noise barriers had been erected. At the same time, measurements were taken at the centre of the motorway, so that any changes in the strength of the source of lead could be taken into account.

28. The monitoring site was typical of situations frequently found alongside motorways in the Birmingham area: the carriageway was on an embankment with the barrier nearby and the houses were below the level of the motorway.

---

*This double glazing was of the type intended to reduce thermal losses from windows rather than that having mechanical ventilation, as provided under the Land Compensation Act.
†Work carried out by the Transport and Road Research Laboratory.

The sampling point for airborne lead was 23·3 metres from the nearest traffic lane, the barrier was 2·7 metres high and the embankment itself was 3·8 metres high.

29. The average weekly concentration of airborne lead measured at the houses was 34·2 per cent of the concentration at the central reservation with the barrier in position, 41·3 per cent without the barrier, and 38·6 per cent with a replacement barrier of a slightly different type. Analysis of the data showed that this difference was not statistically significant: barriers cannot therefore be expected to be useful or protective in the context of airborne lead pollution from major roads, at least under these conditions.

**Lead in dust\***
30. A survey of lead concentrations in dust samples has been conducted for a number of years and more than 5,000 samples have been analysed.† Dust was taken for analysis from a number of different locations, including road surfaces, pavements, gutters, school playgrounds, houses (both inside and outside), and from some commercial premises. Particular attention was given to an area within 400 m of a lead battery factory in order to discover whether local contamination was taking place. All the analyses were performed by atomic absorption spectrophotometry in the Scientific Officer's laboratory. The results of the investigation are shown in Table 9. Median‡ lead concentrations were higher in the dust collected immediately adjacent to the battery factory than elsewhere in the City, although the range of values in the dust collected from roads around the factory was actually smaller than that found in the dust from roads in other parts of the City (450–19,600 ppm compared with 160–50,000 ppm respectively). Conversely, a greater range of values was found in the samples of internal dust collected immediately adjacent to the factory (136–470,000 ppm as against 100–280,000 ppm for the dust from the remainder of the City).

31. The effect which the factory had on local lead concentrations is seen most clearly by considering the proportion of samples with a concentration of 5,000 ppm (0·5 per cent) or more (Figure 8).§ Whereas only 4·5 per cent of samples collected from external surfaces, other than roads, throughout the City exceeded this value, the proportion near the factory was 30 per cent. Similarly, the proportion of samples in excess of 5,000 ppm collected from

---

\*Work carried out by the City of Birmingham Environmental Department.
†Some of the results of this investigation have been reported elsewhere (Archer and Barratt, 1976a and b).
‡The median concentration is the central value of the whole distribution of values: the median is more representative than the mean of a group of results, which may be weighted towards high values by the existence of a small number of samples with exceptionally high lead concentrations.
§This cut-off point was chosen because at the time, this was the maximum amount of lead permitted in the paint for use on children's toys, and it was supposed that young children would be the most likely section of the population to be at risk from lead-rich dust. The legal maximum is now 2,500 ppm. (Statutory Instrument 1974, No. 1367.)

inside houses was much higher close to the factory than elsewhere (24·2 per cent compared with 8·0 per cent). The indications are that the lead does not come from emissions to the atmosphere, but rather that it is physically transported outside the factory on the clothing of the workmen (see paragraph 38 below).

32. From Table 10 it can be seen that the median concentration of lead in dust taken from gutters is directly related to the traffic flow, as would be expected. Thus, the median concentration in samples taken from major arterial (class 'A') roads is 1,800 ppm, whereas on heavily used subsidiary (class 'B') roads the median is 1,300 ppm, whilst on residential roads the median is 950 ppm.

33. Dust samples from pavements in residential areas of the City showed a median concentration of 700 ppm. The concentration inside houses, and in particular from floor coverings, was 950 ppm (median), a value comparable with the concentration in the gutters outside, indicating a considerable degree of physical contamination within the home. This is consistent with the fact that airborne lead concentrations inside the home are comparable with those outside.

34. At the Gravelly Hill interchange itself, 50 individual sites have been monitored regularly at intervals of three months, and during the period April 1974–January 1976, median lead concentrations of 700–900 ppm were found. It was not clear whether there was any seasonal fluctuation, but it is safe to say that dust lead concentrations in the areas of Gravelly Hill are not in any way exceptional for the City as a whole.

35. A direct comparison of these results with those from other cities cannot easily be made because sampling methods and analytical techniques differ. For example, Day and his colleagues (1975) reported a median lead concentration of 970 ppm in the dust of Manchester, a figure somewhat lower than found in Birmingham. In Glasgow a mean of 960 ppm has been found (Farmer and Lyon, 1977) and in London, means from 920–1,840 ppm, depending on the area, have been observed (Duggan and Williams, 1977). It would be unwise to conclude from this, however, that Manchester and Glasgow are less polluted than Birmingham since the pre-treatment of the samples by the three authorities was different. In Manchester and Glasgow the samples were dried and homogenised by grinding and portions of the homogenate were then sent for analysis. In Birmingham the samples were dried and then fractionated, only the smaller particles being used for analysis. Since lead concentrations increase with decreasing particle size (Cerquiglini-Monteriolo and Funiciello, 1972), the treatment for analysis in Birmingham will tend to produce higher lead concentrations than the practice in Manchester or Glasgow.

36. Nevertheless, it is reasonable to conclude that Birmingham as a whole is not unusual with respect to the air and dust lead concentrations by com-

10

parison with other large cities, both in this country and abroad (US National Academy of Science, 1972; Branquinho, C. L. and Robinson, V. J., 1976; Fisher, A. and Leroy, P., 1975; Colacino, M. and Lavagnini, A., 1974; Japan Environment Summary, 1976).

37. It is still not certain whether these concentrations of lead in dust present a hazard to health, and one object of mounting the blood lead study in the pre-school children (described in Chapter 5) was to study this point further, since very young children are the most likely to be potentially at risk because they are more liable to suck contaminated playthings and their fingers.

**The transport of lead on clothing***
38. The results of the dust lead survey showed very clearly that the area around the lead battery factory was much more heavily contaminated than other areas of the City. An investigation conducted by Warren Spring Laboratory for the Central Unit on Environmental Pollution had already indicated that the factory contributed much less to local atmospheric lead concentrations than motor traffic so it was thought that the lead might be leaving the factory on the clothing of the workers, or on vehicles entering and leaving the factory. Both mechanisms would account for the high concentrations in the external dust samples, and the former for the concentrations in the dust taken from some of the nearby houses.

39. To test this hypothesis, certain of the workers were issued with nylon socks to wear for two days (21 and 22 January 1975) so that the amount of lead picked up in that time could be estimated. Nylon socks were chosen since they were cheap, relatively easy to extract with acid and the wearer was in no way conspicuous. With the agreement of the management the socks were issued by the work's Safety Officer to volunteers in the following groups:
  1. administrative staff,
  2. supervisory staff,
  3. manual staff working inside the factory, and
  4. manual staff working outside the factory.
The description of these groups was based on information which the men themselves gave about their work. Each man was asked to work in his normal manner for the two days, and to change his working clothes as he normally would. As a control, eight people who were not employed at the factory and who did not work or visit in the vicinity of the factory wore similar socks on the same two days.

40. Following the exposure period, one sock of each pair was used for analysis, a random check having shown that both socks of the pair were contaminated approximately equally. The lead was removed by extraction with dilute nitric acid and the analysis was performed on the extract by atomic

*Work carried out by the City of Birmingham Environmental Department.

11

absorption spectrophotometry. An unworn pair of socks was also analysed to get a 'blank' reading.

41. The amount of lead on the socks varied with expectation (see Table 11). That is, the manual workers who were directly engaged in handling lead picked up most, whilst the administrative staff in their offices picked up least. The control socks contained more lead than the blanks, showing that some lead had been picked up from the general environment, presumably having been introduced by motor traffic. This amount was very much less than that picked up in the factory however, even from the areas where lead would not be expected to be present to any great extent.

42. Considering that socks are relatively sheltered items of clothing, and that the duration of the experiment was short, the amount of lead picked up was surprisingly high. If other items of clothing were contaminated to a like degree then a considerable amount of lead could be transported from factory to home.

43. After the survey was completed, some further experiments were carried out to study how well lead was removed from socks by ordinary methods of washing. It was found that the lead was not all removed and moreover, there was a transfer of lead from the socks to other clothes which were being washed at the same time. This secondary contamination could prove an additional source of exposure to young children.

44. Concurrently with this survey, a re-appraisal of the hygiene arrangements at the factory was made in an attempt to cut down the transfer of lead from the factory to the general environment. The Health and Safety Executive is revising existing legislation on lead by rationalising all existing regulations relevant to occupational exposure into a single comprehensive set, with the backing of a Code of Practice or Guidance Note. Many of the procedures proposed will be of general application in limiting the transport of other toxic particulate materials from the workplace to workers' homes.

# CHAPTER 3

## BLOOD LEAD CONCENTRATIONS IN SCHOOL CHILDREN

45. The blood lead surveys which were initiated by the Working Party were all designed to give as comprehensive a picture as possible of the range of blood lead values in the chosen population and the factors which might affect them. In particular it was hoped to be able to detect any geographical trends in blood lead concentrations across the City and to identify particular groups of individuals who might be exposed to an unusual degree to particular sources of lead.

46. For the purpose of detecting geographical trends, the City was divided into four areas based on two ring roads (see Figure 9). Area A, the inner zone, was that part of the City totally within the middle ring road; B, the mid-zone, that part between the middle and outer ring roads; and C, the outer zone, was that part peripheral to the outer ring. Area D comprised the whole of Sutton Coldfield which had recently been incorporated into the City after local government re-organisation in 1974.

47. It was decided that children between the ages of 8–14 inclusive would be studied in the first instance and that a statistically adequate sample would comprise 20 boys and 20 girls in each age group and from each of the four areas, a total of 1,120 children in all.

48. With the approval of the Education Authority, schools were chosen in each area from which it seemed likely that the required number of children could be drawn for the study. The Chief Education Officer of the City kindly agreed to write to the Head Teachers in each of the chosen schools outlining the aims of the study and Heads were also sent forms which they were asked to give to pupils to take home to their parents.* These forms again outlined the aims of the study and parents were requested to sign a form of consent if they were willing to allow their child to participate and to permit someone to visit them at home in order to complete a short questionnaire. The signed consents were taken back to school and then returned to the co-ordinator of the study who drew up the list of children from whom blood was to be taken.

49. The response was not as great as had been hoped and consent was obtained for only 730 children (Table 12). The response was greatest in areas B and C and was generally less forthcoming for the older children.

*A copy of this form is at Appendix D.

50. Blood samples were taken during January and February 1975 by doctors who visited the schools by arrangement. The blood was taken by venepuncture into plastic bottles containing potassium EDTA as an anticoagulant. Each child was identified only by a code number which had been allocated by the co-ordinator. The blood samples were identified only by this code number and sent to the laboratory of the City Scientific Officer for analysis. Because there is a great variation in the results of blood lead analyses from different laboratories (Berlin, del Castilho and Smeets, 1972) one in three of the children, chosen at random, had a duplicate sample of blood sent for analysis to the Institute of Child Health in London. (The results of the duplicate analyses appear in Appendix E).

51. The blood was analysed by atomic absorption spectrophotometry using the carbon furnace filter paper method, a modified version of one described by Cernik (1974), in a laboratory which had been established especially for this work.

52. The questionnaires, which were designed to obtain information on possible sources of lead exposure, were completed by health visitors who visited the childrens' parents in their homes. (The questionnaire appears in Appendix F).

## Results

53. Table 13 shows the mean blood lead concentrations, by sex, for the children in each of the four areas. There is no appreciable difference in the overall means obtained for the boys and girls but there is a statistically significant downward trend in mean values from A to D: this is best seen in Figure 10.

54. It should be noted that the numbers in this and succeeding tables do not correspond exactly with those in Table 12. On the day appointed for blood sampling some children were inexplicably away from school, but others apparently stepped in to make up the numbers! This occurred in only a very few cases (4) and since the range of blood lead values was not great (as may be seen in Table 13 and Figure 11) and since values did not vary greatly from one area to another, it is unlikely that this unplanned variation will have influenced the overall pattern of the results.

55. There was no significant variation of mean blood lead concentration with age so far as the boys were concerned, but in the case of the girls the opposite was the case (Table 14 and Figure 12). The mean concentrations were similar in the boys and girls until about the age of 12, after which the girls had a significantly lower value. It is well known that adult females have blood lead concentrations which are consistently 3–5 $\mu$g/100 ml lower than in adult males, and it may be that we are witnessing here the start of this phenomenon (for further discussion on this point see paragraph 75).

14

56. The distribution of children of different ethnic origin varied greatly in each of the four areas (Table 15). In area B the proportion of children of non-Caucasian parents was over 40 per cent and in area A it was approximately a quarter, whereas in C the proportion was less than 10 per cent and in D less than 1 per cent. There was no marked difference between the mean blood concentrations in the different racial groups, however, as may be seen in Table 16. Thus the variation seen in Table 13 is not to be explained on the basis of the different population mix in the four areas of the City.

57. When examining the relationship between blood lead and age of house, a significantly positive upward trend was noted (Table 17), confirming earlier observations (Barltrop and Killala, 1969). Not unexpectedly, inner areas A and B show a higher proportion of children living in houses older than 50 years (22·8 per cent in A, 35·8 per cent in B) than do the outer areas C and D (18·4 per cent and 11·8 per cent respectively). Older houses are thought to be richer sources of lead than modern houses because they are more likely to have lead water pipes or to have old, lead based paint on walls or woodwork. In the present study the presence of lead water pipes was very much more common in older houses, but this does not account for the increased blood lead concentrations. From Table 18 it can be seen that there was no significant difference in blood lead concentrations in children living in houses of equal age irrespective of whether lead pipes were present or not.

58. No environmental studies were carried out in the houses in which the children lived so there is no way of knowing in detail to what sources of lead they were exposed. Children in this age group would not be expected to ingest lead laden dust or flakes of old, leaded paint, as younger children might. However, the preparation of painted surfaces by sanding may produce large amounts of airborne dust, which may be inhaled. There is no information on the proportion of atmospheric lead deposited in and absorbed from the lungs of children – although such data exist for adults – and this is a serious limitation on assessing total exposure.

59. None of the children examined had a blood lead concentration greater than 35 $\mu$g/100 ml, the upper limit of normality indicated in the EEC Directive on Biological Screening of the Population for Lead, and from Figure 11 it may readily be seen that the group as a whole meets by a wide margin the other reference criteria of (a) not more than 10 per cent to have a blood lead concentration in excess of 30 $\mu$g/100 ml; and (b) not more than 50 per cent to have a blood lead concentration in excess of 20 $\mu$g/100 ml. Only 3 per cent of the total group had blood leads in excess of 25 $\mu$g/100 ml, the lowest level at which it has been claimed that untoward effects may occur (David et al, 1972). By far the majority of the children whose blood leads were greater than 25 $\mu$g/100 ml lived in areas A and B as may be seen from Table 19. The actual values for these children, together with their ages are set out in

15

Table 20. The highest value recorded in the survey was 34 $\mu$g/100 ml, found in an eleven year old West Indian girl in area A.

60. Twenty-two of the children had fathers whose occupation involved some contact with lead, but in none of them was the blood lead exceptional, which is gratifying in view of the earlier findings on transportation of lead in clothing (Chapter 2, paragraph 38) and a report of lead poisoning in the children of American lead workers (Baker et al, 1977). The details of the findings in these children are listed in Table 21, together with the occupations of the fathers.

61. Regression analysis of the data indicated that the age of the child's house was the most significant variable in determining blood lead concentrations, with children from older houses having the higher values. Nevertheless, children from area A had significantly higher blood leads than those in other areas, whilst those in area B were affected to a lesser extent. The age and sex of the child also appeared in the regression equation, older girls having significantly lower blood lead concentrations than boys of the same age, or younger girls.

**Conclusions**

62. From the results of this survey it seems reasonable to conclude that the school children in Birmingham are not exposed to lead from any source to an unusual degree. The small range of blood lead concentrations indicates that lead exposure across the City does not vary widely in school children, although it is rather higher (though not unacceptably so) in the central areas than in the periphery. On this evidence, there is no reason to suppose that there is any general risk to the health of children of school age from such exposure to lead. A comparison with the results of other studies suggests that Birmingham school children do not differ to any marked degree from children in other cities with respect to their blood lead concentration (Department of the Environment, 1974; Day, Evans and Robson, 1977).

63. The degree of environmental lead pollution in the City is clearly much less than in some of the large north American cities, judging from the results of blood lead surveys which have been reported. Early results from America indicated that at least a fifth of children had blood lead concentrations in excess of 40 $\mu$g/100 ml (see Table 40 of Waldron and Stöfen, 1974) and even in more recent surveys, approximately 10 per cent of values are over 30 $\mu$g/100 ml (Morbidity and Mortality Weekly Report, 1976).

64. A study of lead concentrations in deciduous teeth from Birmingham children has shown that the mean concentration is similar to that found in children living in suburban or rural areas, and this lends support to the view that children in the City are exposed only to relatively low concentrations of lead (Mackie et al, 1977).

16

# CHAPTER 4

## BLOOD LEAD CONCENTRATIONS IN ADULTS

65. To obtain base-line data on blood lead concentrations in adults, the Working Party considered that the most practicable approach would be to obtain samples from blood donors. Accordingly arrangements were made with the Regional Director of the National Blood Transfusion Service for the collection of samples from donors attending sessions in the City centre.

66. It was hoped to obtain samples from at least 200 men and 200 women, and this necessitated attendance at nine evening sessions during September and October 1975.

67. A leaflet asking for help with the study was prepared (Appendix G) and given to each person attending the sessions. For each of those who agreed to give a sample of blood a slightly modified version of the questionnaire in Appendix F was completed. Those persons who lived outside the City were excluded from the survey.

68. Blood was taken into plastic bottles containing potassium EDTA and identified by a code number; no samples were taken for duplicate analysis. The blood lead concentration was determined by the City Scientific Officer as before (see paragraph 51).

### Results
69. Blood samples were obtained from 225 men and 221 women who were not exposed to lead in their present occupation. A further 9 samples were obtained from men who worked with lead in one way or another; their results were excluded from the main statistical analysis.

70. The mean blood lead concentrations are shown in Table 22. As expected, the mean in the men is higher than in the women and both values are similar to those reported in other countries (Goldwater and Hoover, 1967).

71. Figure 13 shows the percentage distribution of the results. For the females, the distribution and the range (3–35 $\mu$g/100 ml) is similar to that in the school children but for the men there is a wide range (9–46 $\mu$g/100 ml) and the distribution is skewed towards the higher values. For the females, the reference concentrations in the EEC Directive on Biological Screening of the Population for Lead are met with a wide margin; for males, the reference concentrations are not met, although the degree to which they are exceeded is very small.

17

72. Nine of the men and one of the women had blood leads equal to or greater than 35 $\mu$g/100 ml. Details of these individuals are given in Table 23. No follow-up studies were made to identify any potential sources of lead, although four of the men lived in houses with lead pipes. Whether this would have significantly affected the blood lead concentration in these men is open to question, however, since it did not do so for the men (or women) in general (Table 24), or for school children (Table 18).

73. However, Table 24 also indicates that the age of the house did significantly influence blood lead concentrations in both men and women, higher concentrations being seen in the older properties. This is consistent with the results observed for school children (paragraph 57 and Table 17).

74. For the men, blood lead concentration appeared to be unrelated to age, but in the women the concentration increased with increasing age, so that in the oldest age-group the mean concentration was approaching close to that of the men (Table 25). This is seen also in Figure 14 where mean blood lead values are plotted for individual years.

75. We noted earlier (paragraph 55) that school girls' blood lead concentrations were similar to those of the boys until the age of 12, when they began to fall. The concentrations in the adult females remained substantially lower than in the adult males until at least 40, after which they approach the male value. There has never been a satisfactory explanation for the sex difference in blood lead concentrations but we may speculate that it is related to menstruation, since, in the present study, the difference was most marked during the years at which menstruation is taking place.

76. Blood lead concentrations were slightly higher in smokers than in non-smokers, although in women the difference was only evident in those who smoked more than twenty cigarettes a day (Table 26). Almost 64 per cent of the total sample were non-smokers; this proportion, which is higher than would be expected in a sample representative of adults generally, reflects the bias inherent in selection from blood donors and is consistent with the proportion (65 per cent) coming from area C (Table 27). Very few immigrant adults were included in the sample.

77. The uneven spread of the sample makes analysis of areal trends (as was shown to be important for children) difficult. Table 28 indicates no influence of area on blood lead concentrations for men or women, but these results must be treated with caution.

78. Regression analysis indicates that sex is the most important variable determining blood lead concentrations in those not occupationally exposed, whilst age is also important for women. The age of the house was the third variable appearing in the regression equation and smoking the fourth.

79. The blood lead concentrations in the lead workers are given in Table 29; all are equal to or greater than 35 $\mu$g/100 ml and the mean value (45·7 $\mu$g/100 ml) is more than twice the mean of the men who are not occupationally exposed. It is unlikely that any of these concentrations would be considered unduly high for lead workers.

## Conclusions

80. Blood lead concentrations in adults in Birmingham are similar to those reported from other cities (World Health Organization, 1977; Day et al, 1977), which tends to confirm the view that the general level of environmental contamination in the City is not great. A little caution is needed here, since the sample was clearly not representative of the adult City population in general, either geographically or ethnically.

81. The range and distribution of blood lead concentrations in women are similar to those in the school children, but in the men, there is a slight shift in the distribution towards higher values, even when lead workers are excluded, which suggests that some men are exposed to local sources of lead which are not encountered generally. We do not know what these local sources are.

# CHAPTER 5

## BLOOD LEAD CONCENTRATIONS IN PRE-SCHOOL CHILDREN

82. Although the results of the two blood lead surveys described in the preceding chapters were reassuring, the Working Party decided that a further survey of blood lead concentrations in pre-school children should be undertaken.

83. This was necessary for two reasons. First, because exposure of the 1–5 age group is not necessarily related to exposure of adults and older children in the same areas (for example, Sayre and his colleagues (1974) found a mean blood lead in New York adults of 20 $\mu$g/100 ml, but a mean in under fives of 44 $\mu$g/100 ml; and Singh et al (1975) showed that a third of pre-school children in Newark, New Jersey, had blood leads greater than 40 $\mu$g/100 ml); and second, because blood lead concentrations tend to be maximal between the ages of 1 and 3, and it is in this age group that most cases of lead poisoning occur (Chisholm and Harrison, 1956).

84. Having obtained the sanction of the Research Ethical Committee of the Central Birmingham Health District, it was proposed to take samples from 20 boys and 20 girls aged 1–4 in each of the four areas defined in Chapter 3, a total of 640 samples in all. Since it was especially important to define the true extent of the problem – if indeed there was one – it was decided to draw a random sample from the birth records of the City.

85. To allow for a possible poor response rate, 1,600 names were drawn at random from the records, making due allowance for the different ethnic mix in the various parts of the City. Having alerted local General Practitioners through the local Medical Committee, letters were sent to the parents of each of the children in the sample, asking if they would be willing to participate in the survey.* As may be seen from Table 30, however, the response was not nearly as good as had been expected. More than half those to whom letters were sent gave no reply at all. An attempt was made to improve the response by publicity through the local press and radio and by employing a person to visit the houses from which there had been no reply. This was a singularly unrewarding exercise, however, since less than 20 additional participants were obtained.

86. All those who agreed to help were given an appointment to bring their child to a ward at Good Hope Hospital, Sutton Coldfield, which had been set

---

*A copy of this letter is at Appendix H.

20

aside especially for our use. Transport to and from the hospital to home was provided for those requiring it, in the hope of maintaining an adequate response rate.

87.  Capillary blood samples were taken from the finger in March 1977 by two technicians who were instructed in the technique at the Institute of Child Health in London in order to minimise the formidable problem of skin contamination.

88.  The blood sample was taken into heparinised tubes, identified only by a code number for analysis by the City Scientific Officer; duplicate samples were taken in a fifth of cases, chosen randomly, for duplicate analysis at the Institute of Child Health. At the same time, the accompanying parent was interviewed in order to complete the standard questionnaire to which the only significant addition was a question designed to see whether or not the child had pica.*

89.  When the survey began, the local press, radio and television gave it some publicity, highlighting the difficulties which there had been in getting sufficient numbers. Many parents who came to hear of the rather poor initial response in this way offered to bring children who had not been selected in the original sample; some of these parents already had one child in the sample, but most did not. None who came was refused (even if, as rarely happened, the child offered was considerably over-age), but a note was made to show that they were not part of the random sample.

**Results**
90.  There was a tremendous variation in response from the various parts of the City as may be seen from Table 31, which gives the numbers of attendances by area. Bearing in mind the aim of examining 160 children from each area (20 boys and 20 girls in each age group), it can be seen that the response from the inner area of the City was extremely low. About 20 per cent of parents in area B, nearly half in C, and over three quarters in Sutton Coldfield responded favourably. There is no doubt that the fact that the blood samples were taken at Sutton Coldfield was a factor in motivating the parents to respond as they did: had we been able to arrange for blood to be taken more centrally as we had originally intended, then it is possible that parents living in the centre of the City would have been more willing to co-operate.

91.  What was particularly disappointing was that over a hundred of those who had initially agreed to help did not keep their appointments; thus of the total of 353 (randomly selected) who agreed to help (Table 30), only 243 kept the appointment (Table 31).

---

*Pica is the habit of eating or mouthing non-food materials.

92. The number of blood lead results finally obtained is less than shown in Table 31, since in some cases it was not possible to take a sample and in a few cases, some misadventure overtook the sample on the journey from hospital to laboratory. In all, a total of 429 blood samples were analysed of which approximately 55 per cent were from the random sample. The results are given in Table 32. The mean values are higher than in the school children and the range is considerably wider (Figures 15–17); the means are also generally higher for the non-random than for the random sample as Einbrodt et al (1975) also found. As with the school children, mean concentrations were lower in areas C and D than in area B; the data for area A are too few to interpret.

93. By contrast with the school children, fifteen of the under fives had a blood lead concentration equal to or greater than 35 $\mu$g/100 ml (Table 33). The high number of children with raised blood leads in area B, considering the small size of the total sample from this area, was particularly disturbing. Table 33 and Figures 15–17 indicate that the reference blood lead distribution in the EEC Directive on Biological Screening of the Population for Lead is met at the 50 per cent level (20 $\mu$g/100 ml) but exceeded (albeit by a small margin) at the 98 per cent level (35 $\mu$g/100 ml), mainly in the case of boys.

94. There were, in addition to these 15 children, 2 over-age children who also had high blood leads. All seventeen children were followed up in order to obtain further blood samples (see paragraph 98 below).

95. The variation in blood lead concentration with age is shown in Table 34 and in Figure 18. The mean values do not vary much, although there is generally a slight peak at the age of 2 or 3. Variation with age of house (Table 35) shows that those children living in the newest houses have the lowest blood leads, but otherwise, there is no consistent trend with age of house. The presence of lead water pipes in the house apparently has very little effect on blood leads (Table 36), and although this finding needs to be treated with caution in view of the large number of "don't knows" in the sample, it is clear from the small deviation from the mean values that the finding could not be significantly altered by a better knowledge on this point in this particular sample.

96. One rather surprising feature was that there was no correlation between blood lead concentrations and a history of pica (Table 37 a–c); this is the route by which young children showing evidence of enhanced exposure are usually found or presumed to take in lead from their surroundings.

97. Only a small number of the children had fathers whose occupation involved occasional contact with lead, and all were painters and decorators. The blood lead concentrations of these children are shown in Table 38; none

is exceptional. This is perhaps not surprising in view of the reduced use of lead in paints in recent years.

**Follow-up studies**

98. The parents of all the children whose blood lead concentrations were 35 $\mu$g/100 ml or over were asked to make an appointment for a further blood sample to be taken. All the second samples were taken in the home and all in duplicate. The results are shown in Table 39. In some cases, the blood lead level was lower on the second occasion than on the first, but ten children were still reported by one or other of the laboratories as having a level of 35 $\mu$g/100 ml or greater. Agreement between the two laboratories was good except in one instance (the last child in Table 39) where the result obtained by the City Laboratory was twice that reported by the Institute of Child Health. (See Appendix E for a consideration of analytical quality control.)

99. None of the fathers of the children involved worked with lead and in only one case was the house in a poor state of repair. One child was not included in the follow-up study because the house in which he was thought to be living was derelict and boarded up. The child had been selected in the random sample and had been brought to the hospital by his mother, but we have no reliable information as to where he was living at the time.

100. Since the second series of blood samples was taken, one child has returned to Pakistan. The nine other children with enhanced exposure were all referred to their family doctor by the Area Specialist in Community Medicine (Environmental Health) (SCM (EH)). These children were subsequently seen by a consultant paediatrician who is a specialist in neurological diseases. There were no frank significant neurological findings; minor abnormalities were seen in three children but inevitably these could not be specifically related to enhanced exposure to lead as they could equally have been related to other factors. The School Health Service of Birmingham Area Health Authority, in conjunction with the family doctors, will monitor the future development of the nine children with blood lead concentrations in excess of 35 $\mu$g/100 ml.

101. The SCM (EH) has written to the parents of the other six resampled children to tell them that their latest blood lead concentrations are satisfactory and that no further action will be taken.

102. An environmental assessment was made of each house where a child with raised blood lead concentrations was found; details are given in Appendix I. To summarise, the work included sampling and analysis of internal and external dusts, soils, paint flakes and water, and an examination of a number of other factors thought relevant; these included the habit of pica, the presence of flaking paint, evidence of recent redecoration, the proximity of the home to a major road or works using lead or lead compounds, the proximity of such works to the parents' places of employment, the period of resi-

dence at the current home address, the parents' occupations, the growing of garden vegetables and the possible use of lead-containing cooking utensils, culinary powders, cosmetics and medicines.

103. Most of the above factors were not implicated as causes of enhanced exposure. In one house it proved impossible to attribute the raised blood lead concentrations to any factor and in five other cases only isolated instances of high concentrations of lead in dust or in paint flakes were seen. Only in a single house was the source of undue exposure found with some reliability; here extensive contamination with lead-rich dust, derived from the preparation of old painted surfaces, together with the habit of pica in the children, seems the likely cause. This may indicate that a general problem exists where dust-producing methods of paint preparation are used in old houses where children with pica are living.

**Conclusions**
104. The survey amongst the children of pre-school age revealed a relatively high proportion of blood lead concentrations of 35 $\mu$g/100 ml or above. Amongst the children living in the middle zone of the City the proportion was particularly high. Of these, children of parents of Asian origin contributed the bulk of cases. The response from the central inner area was inadequate to assess the distribution of exposure.

105. No frank neurological impairment was seen in those children with raised blood lead concentrations; minor abnormalities could inevitably not be attributed to exposure to lead.

106. In only one case has an environmental assessment of the child's home revealed with some reliability the likely cause of undue exposure. Recent redecoration of old painted surfaces, leading to extensive contamination by lead-rich dust, allied with the habit of pica, seemed to be the likely cause. Respiration of airborne lead from traffic has not been found to be important.

107. There is a clear need to assess with more certainty to what extent pre-school children generally living in the centre of the City have high blood leads. The yield from a postal approach was extremely disappointing, perhaps because the hospital at which blood was taken was too remote; or perhaps because many of the immigrant families did not understand fully the contents of the letters which were sent to them. The results of this study have shown that a problem exists; its scale will not be known until we have more data, and an extension of this work is being organised.

# CHAPTER 6

## SUMMARY AND CONCLUSIONS

**Atmospheric lead and dust**

108. The opening of the Gravelly Hill interchange was accompanied by a considerable increase in motor traffic but atmospheric sampling in the area has shown that there has not been a concommitant increase in the air lead concentration. Monthly mean atmospheric lead concentrations at the site nearest the centre of the interchange are generally less than 3 $\mu$g/m$^3$, and on an annual basis average concentrations at this site are less than 2 $\mu$g/m$^3$. At sites further away from the interchange, annual mean values are about 1 $\mu$g/m$^3$.

109. In suburban sites away from heavy traffic it has been found that the atmospheric lead content is well below 1 $\mu$g/m$^3$, so that undoubtedly the presence of the motorway interchange results in an increase in the local concentrations. The elevated design of the interchange and the open nature of the area make possible rapid and widespread dispersion of the lead emitted from the petrol-driven vehicles using it; this is probably the reason why local atmospheric concentrations are not higher. Some support for this suggestion comes from the observation that wind speed is the most important meteorological factor in determining airborne ambient lead concentrations.

110. An investigation of three houses showed that airborne lead concentrations inside houses vary similarly with those outside and follow the same diurnal variation. Double-glazing (introduced as a noise-reduction measure) appears to offer no protection in this respect.

111. Dust gathered from roads throughout the City shows a median lead concentration of 1,200 parts per million; the median concentration in dust from other external surfaces is also 1,200 ppm, whilst internal samples show a median concentration of 1,000 ppm. In the vicinity of a large battery factory concentrations are considerably higher and the proportion of samples containing more than 5,000 ppm is also greater.

112. A preliminary investigation showed that employees at the factory were taking lead out of the factory on their clothing, and concurrently with this study, hygiene arrangements at the factory were altered.

**Blood lead concentrations in school children**

113. The survey of blood lead concentrations in school children (aged 8–14)

gave reassuring results in that mean values were low (15·0 $\mu$g/100 ml in boys and 14·2 $\mu$g/100 ml in girls) and no child was found with a blood lead greater than 35 $\mu$g/100 ml.

114. Whilst the range of blood lead concentrations found in the school children was relatively small, suggesting that exposure to lead was fairly uniform, children from the inner city areas had slightly higher blood lead concentrations than those from the outlying districts.

115. Blood lead concentrations were positively associated with the age of the house in which the child lived, but the reasons for this are not known at present. It did not appear to be due to the presence of lead plumbing. This is another topic warranting a more detailed study.

116. Girls generally had lower blood lead concentrations than boys, and concentrations in girls decreased with increasing age.

117. About one third of the blood samples were taken in duplicate for analysis at the laboratory of the Institute of Child Health in London. The agreement between the results from this laboratory and those from the laboratory of the City's Scientific Officer, where all the other blood samples were analysed, was extremely close.

**Blood lead concentrations in adults**
118. In adults, mean blood lead concentrations were found to be 22·1 $\mu$g/100 ml in men and 14·7 $\mu$g/100 ml in women, values consistent with those reported elsewhere. The range in the women was similar to that in the school children, but in the men the distribution was skewed towards higher values, nine being 35 $\mu$g/100 ml or above, whereas only one woman had a value of 35 $\mu$g/100 ml. In none of these cases was an unusual source of lead exposure evident.

119. Blood lead concentrations in women increased with increasing age, to meet the values in men at ages in excess of about 40. As with school children, blood lead concentrations were higher in people living in older houses. Smokers had higher blood lead concentrations than non-smokers. Some caution must be exercised in interpreting these results, since the adult sample population was clearly unrepresentative.

120. The adults from whom the samples were obtained were all blood donors and amongst those we studied we found nine lead workers. All nine had blood lead concentrations equal to or greater than 35 $\mu$g/100 ml, the highest being 62 $\mu$g/100 ml; this would not be considered unduly high for a lead worker.

**Blood lead concentrations in pre-school children**
121. The Working Party proposed that pre-school children (aged 1–4) should

be sampled at random so that as accurate a picture as possible would be gained of the blood lead distribution in this group, which is the one most likely to be at risk from environmental lead. The response from the randomly selected group, however, was very disappointing, particularly from the inner City area where only about 5 per cent of those approached agreed to participate in the study. In the event, many children who had not been selected were brought to the clinic where the blood was being taken, and samples were also taken from these children.

122. Overall, the mean lead concentration in the blood of the pre-school children was 20·1 µg/100 ml and the distribution was similar to that of the adult males. The mean for the randomly chosen children was 18·8 µg/100 ml whereas in the remaining non-random children it was 21·6 µg/100 ml.

123. However, a disturbing feature of this part of the work was the relatively high number of children (15 out of 429) with concentrations equal to or greater than 35 µg/100 ml. Seven of these children were from the two inner areas of the City, an unacceptably high proportion when one considers that only 83 children in all were examined from these areas. All but one of these children were of Asian origin and there was in addition another Asian child of 6 whose blood lead was also over 35 µg/100 ml.

124. Repeat blood samples were taken from the children with high lead concentrations and in ten the value was confirmed as being over 35 µg/100 ml. One child has since returned to Pakistan. The other nine were referred to their family doctors and were then seen by a consultant paediatrician. None showed frank signs of significant neurological impairment and minor abnormalities unattributable to specific causes were observed in only three. The Environmental Health Department of the City has carried out an environmental survey of all the children's homes but in only one case has the probable cause of undue exposure been found. Here, extensive contamination with lead-rich dust, derived from preparation of old painted surfaces, together with the habit of pica, seems to be responsible.

125. There is clearly a need to establish whether or not there is a general problem of exposure to lead for young children living in the central areas of the City, to determine its scale and to identify the sources of lead. Work has already been put in hand to this end.

**The future of the Working Party**
126. The Joint Working Party has fulfilled its brief. We have found that the degree of environmental contamination around Gravelly Hill, though it increased when the interchange opened, is not markedly greater than elsewhere in the City; we have also established that the distributions of blood lead concentrations of school children and adults living in the area are typical of urban dwellers.

127. However, our latest work does suggest that a relatively high proportion of very young children living in the inner City areas have elevated blood lead concentrations. We have already indicated that this is a matter which urgently requires further investigation. The precise reasons for this finding are not yet known but the problem is clearly not related to the motorway interchange itself, nor indeed does it seem at all likely that airborne lead in general is directly responsible. The work is now entering a new phase and it is no longer sensible to treat it as an off-shoot of the particular points of public concern from which we started.

128. The success of this new phase will, however, continue to depend upon the co-ordination of local and central agencies and we recommend that a small Steering Committee, comprising the relevant local and central interests, be established to supervise the project. This Steering Committee should also co-ordinate as necessary other lines of research, such as the need for more work around local industrial sources of lead, and a long-term follow-up of the children from whom we have already taken blood samples.

129. Birmingham is in many ways an ideal centre on which to base studies on environmental lead, since it is an important centre of the non-ferrous metal-working industries and an excellent liaison mechanism between the City, central government and academic institutions has been established. It is hoped that liaison will continue in the form of the proposed new Steering Committee, in order to build upon the considerable experience and expertise in this field which has been developed.

# APPENDIX A

## MEMBERSHIP OF THE JOINT WORKING PARTY ON LEAD POLLUTION AROUND GRAVELLY HILL

*Chairman:*
Regional Director, West Midlands Region, Department of the Environment; from March 1974 to September 1975 – Mr. J. E. Hannigan; from September 1975 to the present – Miss S. W. Fogarty.

*Secretary:*
J. P. Giltrow,* Central Unit on Environmental Pollution, Department of the Environment.

*Members:*
F. J. C. Amos, Chief Executive, City of Birmingham (until July 1977);
Dr. W. Nicol, Area Medical Officer, Birmingham Area Health Authority (Teaching), Birmingham;
E. N. Wakelin, City Environmental Officer, City of Birmingham Environmental Department (until his retirement in 1976);
A. Archer, initially as Assistant City Environmental Officer, Environmental Department, and latterly as Chief Officer, Environmental Department;
Dr. R. S. Barratt, Special Duties Officer (Air Pollution), City of Birmingham Environmental Department (until June 1977);
Dr. H. A. Waldron,† Department of Social Medicine, The Medical School, University of Birmingham;
Dr. R. Stephens, Department of Chemistry, The University of Birmingham;
Dr. J. D. Butler, Department of Chemistry, The University of Aston in Birmingham;
F. S. Alexander, West Midlands Regional Office, Department of the Environment;
D. M. Colwill, Transport and Road Research Laboratory, Department of the Environment (latterly Transport), until April 1977;
Dr. R. St. J. Buxton, Department of Health and Social Security;
Dr. S. Ruttle, Department of Health and Social Security;
Mr. P. Kingslan, Statistics, Planning and Regional Division, Department of the Environment, until July 1976; Mrs. P. Dowdeswell until April 1977; thereafter Mr. P. Tan.

---

*Now with Her Majesty's Alkali and Clean Air Inspectorate, Health and Safety Executive.
†Presently at the TUC Centenary Institute of Occupational Health, London School of Hygiene and Tropical Medicine.

29

The Working Party also wishes to express its appreciation of the help provided by Dr. J. Preston (District Community Physician, Sutton Coldfield), by the Area Administrator, Good Hope Hospital, Sutton Coldfield and by the Nursing Authority in carrying out the work on young children; to Dr. S. S. Bakshi, Specialist in Community Medicine (Environmental Health), Birmingham Area Health Authority (Teaching) for the medical follow up; to Dr. S. H. Green, Senior Lecturer in Paediatrics and Child Health, The Children's Hospital, Birmingham, for carrying out the examination of pre-school children with raised blood lead concentrations; to the Education Authority for help provided in the work on school children; to the City Scientific Officer and the Institute of Child Health for analytical services; to Mrs. C. Barratt of the Environmental Department for much of the organisational work; to Mrs. K. Hough and Miss P. Culbert, who took blood samples from the pre-school children; to Mrs. L. Vickerstaff for help with co-ordinating the medical work; and to the Press and Broadcasting Services for their help in publicising the work.

# APPENDIX B:
## CONVERSION OF BLOOD LEAD CONCENTRATIONS FROM TRADITIONAL TO S.I. UNITS

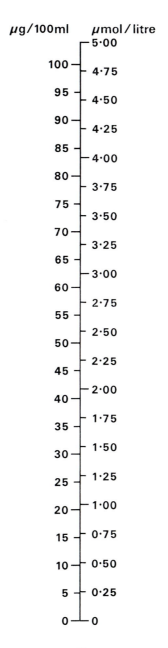

| µg/100ml | µmol/litre |
|---|---|
| | 5·00 |
| 100 | 4·75 |
| 95 | 4·50 |
| 90 | 4·25 |
| 85 | 4·00 |
| 80 | 3·75 |
| 75 | 3·50 |
| 70 | 3·25 |
| 65 | 3·00 |
| 60 | 2·75 |
| 55 | 2·50 |
| 50 | 2·25 |
| 45 | 2·00 |
| 40 | 1·75 |
| 35 | 1·50 |
| 30 | 1·25 |
| 25 | 1·00 |
| 20 | 0·75 |
| 15 | 0·50 |
| 10 | 0·25 |
| 5 | 0 |
| 0 | |

# APPENDIX C

## BLOOD LEAD CONCENTRATIONS IN PEOPLE LIVING NEAR THE M6–A38(M) INTERCHANGE

1. The opening of the M6–A38(M) interchange presented the City of Birmingham with what was probably a unique opportunity, at least in terms of scale, to study the effects of an increase in traffic flow on the blood lead concentrations of local residents. The Aston Expressway (the A38(M)) opened on 1 May 1972 and the M6 on 24 May 1972, and in the eight or nine months which followed, the number of vehicles coming into the district more than doubled. In April 1972, a total of 361,040 vehicles per week was recorded at Salford Circus; in March 1973, the weekly total recorded on the Aston Expressway and Salford Circus had risen to 748,639 of which approximately two thirds were petrol-engined vehicles; in March 1974 the weekly figure was 900,451.

**Study population**
2. The study population comprised a group of adult men and women, and children under ten years of age, all living within 600 m of the centre of the interchange. All the residents living within this area had been approached in March and April by the staff of the Public Health Department and approximately 2,200 agreed to participate. Of these, about 1,600 gave blood when the time came.

3. Three series of blood samples were taken, the first in May 1972, immediately before the interchange opened, the second between October 1972 and March 1973, and the third between October 1973 and January 1974.

4. Only about 900 of the original participants gave blood a second time, whilst for the third series, the number had fallen to 689.

5. For the purposes of statistical analysis, the subjects were allocated into one of three geographical areas as follows:
    1, within 100 m of the centre of the interchange;
    2, between 100 and 300 m of the centre;
    3, between 300 and 600 m of the centre.
No control population was studied.

**Blood sampling and analysis**
6. For the first series, capillary blood samples were taken. Initially blood was taken from the ear lobe but this was abandoned as being too difficult after the first 150 samples, and the remainder were obtained from a thumb

prick. The report of the Senior Administrative Medical Officer for Environmental Services, made to the Health Committee in July 1973, makes clear that a considerable amount of squeezing was used in collection of these earlier samples. This is an important point because squeezing might lead to an unusually high proportion of plasma in the sample (haemo-dilution) and thus (since most lead in blood is associated with the red cells) a mis-leadingly low value for lead in blood. The effect of this is considered later. From thumb prick samples, a number of very high blood lead concentrations were reported, presumably because the skin was contaminated with lead and had not been thoroughly cleaned. When this occurred two further samples were taken and the lowest value obtained has been used in the statistical analyses.

7. Because of the difficulties with skin contamination, venous blood was taken for the second series, although in 100 cases capillary blood was also taken so that a comparison could be made between capillary and venous blood levels. Venous blood only was taken for the third series.

8. All the samples were analysed by atomic absorption spectrophotometry. The first two series were analysed in the Regional Toxicology Laboratory at Dudley Road Hospital, Birmingham, but the third series was analysed in the laboratory of the City Analyst.* One hundred duplicate samples from the third series were analysed in both laboratories.

**Results**
9. So many variables were introduced during the course of the study that it is difficult to draw any meaningful conclusions from the results which were obtained. Nevertheless, it is appropriate to record the results and interpret them insofar as this is possible, with due account being taken of confounding factors. We have disregarded the results of those people in the first series who did not also contribute to the second series. The results from the remainder of the first series are shown in Table 40. As may be seen, the numbers of subjects in each area were not equal, nor were the mean blood lead concentrations equal in each area. The values in the men and in the young children were consistently higher than in the women; however, it should be noted that in areas 1 and 2, the mean values were quite different from those typical of adults in general (Chapter 4) – being substantially lower – whilst for area 3, mean values for women and children were similar to those found elsewhere (Chapters 3 and 4). Such discrepancies make interpretation very difficult.

10. In the second series (Table 41), the mean values had apparently risen in both males and females, so that these data now indicated values comparable with those described earlier for adults. In children, mean values had apparently risen only in area 1 and fallen in areas 2 and 3. It must be remembered

---

*The City Analyst became known as the Scientific Officer after the re-organisation of local government in April 1974. The Regional Toxicology Laboratory is referred to as RTL in what follows.

in drawing these comparisons that a change of sampling technique occurred between series and the effect of this is not known with certainty.

11. However, the effect of changing the sampling technique was studied in the 100 samples of venous and capillary blood taken in Series 2 (paragraph 7) and analysed in the same laboratory (RTL). The capillary mean was 19·5 $\mu$g/100 ml and the venous mean 14·8 $\mu$g/100 ml; the mean difference of 4·7 $\mu$g/100 ml was highly significant, p<0·001. Thus if these pairs of samples were typical of the series as a whole (and this is doubtful since they were not selected at random) then the effect of changing from capillary to venous blood would be to lower the mean by about 5 $\mu$g, so *minimising* the rise in blood lead concentration noted in the second series.

12. There remains the problem of haemo-dilution and the possibility that values in the first series were anomalously low. However, the differences in *paired* capillary samples in the two series (Table 42) were not generally significant for adults, an effect being evident only in area 2. Moreover, the differences in paired capillary samples were not *consistently* greater than the differences in the capillary – venous means of the whole sample of adults (Table 43); data on paired samples of children are too sparse to be meaningful. Again the validity of these results is in doubt since the subjects were not selected at random.

13. The mean values in the third series (Table 44) were apparently considerably higher than in Series 2 (Table 41), but no difference between the three areas in the third series was evident. It would appear that the mean values in all cases were quite different from those observed elsewhere for adults and children (Chapters 3 and 4). It should also be noted that different laboratories were responsible for the analyses shown in Tables 41 and 44. A summary of the changes in all three series is shown in Table 45; this is the type of interpretation which gave rise to the concern in early 1974 that blood lead concentrations locally were increasing rapidly, but it is vital to note that such interpretations ignore the effects of the confounding factors.

14. As indicated above, there is a need to comment upon the complications introduced into the interpretation by a change of laboratory undertaking the analyses. In 100 paired venous samples taken in the third series, the mean reported from the RTL was 21·1 $\mu$g/100 ml and from the City Analyst, 26·4 $\mu$g/100 ml. The mean difference of 5·3 $\mu$g/100 ml was significant at the 0·1 per cent level. Thus, changing the laboratories had the effect of increasing the apparent difference in blood lead concentration between the second and third series. When account is taken of this, the apparent rise is reduced from 6·5–11·8 $\mu$g/100 ml (Table 45) to 1·2–6·5 $\mu$g/100 ml. The overall apparent rise from first to third series is also reduced by a similar amount for the same reason.

15. To overcome the problems associated with the effects of changes in sample groups, we have analysed the blood lead concentrations in those subjects who gave blood to all three series. The results of this analysis are shown in Tables 46a, b and c. A summary of the changes is given in Table 47 and it will be seen that they are broadly similar to those seen in the larger sample (cf Table 45), but it must again be remembered that the comments made previously about the apparently unusual nature of the mean values and about changes in laboratories used remain relevant.

16. To overcome both problems associated with change of sample group and change of laboratory, a small group of subjects was found who not only had blood taken in each of the three series, but all of whose samples were analysed at the RTL. All the subjects were from area 2, i.e. they lived between 100 and 300 m from the centre of the interchange. The apparent rise in blood lead concentration is approximately equal between series 1 and 2, and 2 and 3, and the rise is less than seen in the large group in series 2 (Table 48: cf Tables 45 and 47). All the differences are statistically significant.

17. Finally, within this small group of subjects, those who gave blood in series 2 and 3 represent a group for whom *all* confounding factors can be eliminated, i.e. the same people were sampled in each series, venous samples were taken on each occasion and analysis was carried out at a single laboratory (RTL). It is abundantly clear from Table 48, series 2 and 3, that a statistically significant increase in mean blood lead concentrations occurred in these people in the period October 1972/March 1973 and October 1973/ January 1974.

**Conclusions**
18. There is no doubt that, however one looks at the results, the mean blood lead concentrations of some of the residents living near the M6–A38(M) interchange increased after the motorway opened. How much this was a direct consequence of the increase in traffic flow is difficult to say, because of the way in which the study was carried out; the fact that so few people remain in groups for which a reliable interpretation can be given is evidence of the complexity of surveys of this kind.

19. If the rise in blood lead concentrations were due to an increase in airborne lead concentrations, then it is difficult to understand why the rise should be as great or greater in men than in the women or children because the men would presumably spend a smaller part of their day in or near their home. Moreover, it is also difficult to reconcile the changes in blood lead with data on airborne lead concentrations, because other research has shown (Wells et al, 1977) that an increase in airborne lead concentrations of 1 $\mu g/m^3$ can be expected to lead to an increase in blood lead concentrations of about 0·5–2 $\mu g/100$ ml. Since no increase in airborne lead concentrations took place

between October 1972 and January 1974 (Figure 5), the rise in blood leads in this period cannot be attributed to respiration of airborne lead.

20. The small group of people whose blood samples were all analysed at the Regional Toxicology Laboratory come closest to a controlled group and their mean blood lead concentrations rose 4–5 $\mu$g/100 ml between each series. How typical they are of the residents as a whole, however, cannot be said with any confidence. Thirty-eight of these men and fifty-three of the women again gave blood samples in October 1974. This time the analyses were made in the City's laboratory and the mean values were virtually identical with those which the City Analyst had found in the previous year (Table 49). On this limited evidence, it appears that the blood lead concentrations were no longer rising.

21. Although we cannot be certain that lead emitted from motor vehicles using the new motorway caused the increase in blood lead concentrations which were observed, no alternative hypothesis has yet been put forward.

22. One further point needs emphasising; the mean blood lead concentrations are all well below the values associated with frank disease, although they are higher for women than those which have been found in the more recent studies initiated by the Working Party (Chapters 3, 4 and 5).

# APPENDIX D

## LETTER TO PARENTS OF SCHOOL CHILDREN

January 1975

Dear

I expect that you will have read or heard of the work which is being done in regard to lead pollution. To get the results into perspective, we think that it is most important to measure the levels of lead in the blood of children living within the City boundaries by taking blood from about 1,100 children between the ages of 8 and 14, both boys and girls.

We have chosen some schools for our study and we would very much like you to give your consent for us to take blood from your child. The blood will be taken at the school from a vein in the arm by an experienced health visitor who would also wish to interview you so that she can fill in a simple questionnaire.

The fact that we are taking blood at the chosen schools **does not mean that we think that the children are in any danger from lead whatsoever** and we will let you have the result of the test in due course.

If you are agreeable that your child can take part in this important study, please sign the form of consent below and return the letter to the Head Teacher.

Thank you very much for your help.

Yours sincerely

Councillor John Charlton
(Chairman, Environmental Services Committee)

I hereby give my consent for my child..................................................
to have blood taken for the study conducted by the City of Birmingham.

Signed.............................. Date...........................
Parent/Guardian

Address................................................

................................................................

................................................................

All letters to be addressed to the Chief Officer. Telephone Calls to Direct line

## APPENDIX E

## QUALITY CONTROL IN BLOOD LEAD ANALYSIS

**Inter-laboratory comparisons**

1. To undertake the blood lead analyses, the Scientific Officer of the City of Birmingham (SO) established a special laboratory. So that the results he obtained could be validated, a number of duplicate samples were taken at random from children included in the studies described in Chapters 3 and 5, and sent for analysis to the laboratory of the Institute of Child Health in London (ICH), a centre which has great expertise in this technique.

2. Approximately one in three of the samples from the school children were duplicated and one in five from the children of pre-school age. The complete list of results is shown in Tables 50 and 51. A comparison of the means of the samples and the results of paired $t$-tests are shown in Table 52.

3. The difference in the means of the blood from the school children was 1·8 $\mu$g/100 ml, which is statistically highly significant. In practice, however, the results are as close as might reasonably be expected, especially when one considers the magnitude of the variations found between other laboratories (Berlin et al, 1972). There is a positive correlation between the results ($r=0·74$; $p<0·001$) and the regression equations are:

$$SO=1·21+0·81 \text{ ICH}$$
$$ICH=6·48+0·68 \text{ SO}$$

The regression lines obtained from these equations are shown in Figure 19.

4. For the capillary samples, the difference between the means is a mere 0·1 $\mu$g/100 ml, the correlation coefficient is 0·71 ($p<0·001$) and the regression equations are:

$$SO=3·25+0·84 \text{ ICH}$$
$$ICH=7·58+0·61 \text{ SO}$$

The regression lines are drawn in Figure 20.

**Comparison of venous and capillary blood**

5. Since there are differences in the concentration of lead in venous and capillary blood and since it was necessary to take capillary and not venous blood from the pre-school children, it seemed important to determine the extent of the differences as measured in the City laboratory.

38

6. Thirty-eight adults, with no occupational exposure to lead, volunteered to donate venous and capillary blood samples, which were taken into the same anticoagulants as in the main studies. The results of the analyses on these samples are shown in Table 53 and in Figure 21. In general the concentration in capillary samples is higher than in the venous samples, the mean difference being 2·7 $\mu$g/100 ml, statistically significant at the 5 per cent level. (As in previous work, the mean value in men, 26·8 $\mu$g/100 ml, venous, was higher than in women, 16·7 $\mu$g/100 ml, venous.)

7. There is a highly significant ($p<0·001$) correlation between the two sets of results ($r=0·73$) and the regression equations are:

$$\text{venous} = 4·95 + 0·68 \text{ capillary}$$
$$\text{capillary} = 7·42 + 0·77 \text{ venous}$$

# APPENDIX F

## QUESTIONNAIRE

## CITY OF BIRMINGHAM BLOOD LEAD SURVEY

NORMAL/SPLIT/TRIPLE                                    Code No

Date of Sample:

Name:

Age:                                    Date of birth:

Place of birth:                         EO:                    Sex:

School:

Address:

How long at this address:

If less than 1 year, previous address:

Age of house:                          Lead water pipes:   YES/NO

Vacuum cleaner:   YES/NO

Child's milk intake (pints/day):

Hobbies of those in the house:

   Wine making:   YES/NO          Enamelling:   YES/NO

   Pottery:   YES/NO              Home Decorating:   YES/NO

Occupation and place of work of:

   Father:                        Mother:

   Other sibs.:

Previous medical history (including dates of admission to hospital where relevant):

Parents at home during the day:   YES/NO

Telephone Number:

## INFORMATION SHEET FOR ADULTS

## CITY OF BIRMINGHAM BLOOD LEAD SURVEY

As part of its research programme into lead pollution, the City of Birmingham is trying to establish normal levels of lead in the blood of adults and children living within the City. We are seeking for volunteers to help in what we think is a very important study and we wonder whether you would help us? We are asking that you allow us to take a small portion of the blood which will be taken during your donation so that we may analyse its lead content. We would also like to ask you a few details about your occupation, place of residence and so on, to help us establish how much lead you are likely to come into contact with.

We are trying to establish the NORMAL levels of lead in the blood in this survey so there is no suggestion whatever that you are likely to be suffering in any way from lead poisoning. If you wish, we will send you the results of the test in due course.

We will come and see you during the session to ask if you are willing to help us.

Dr. H. A. Waldron

All letters to be addressed to the Chief Officer. Telephone Calls to Direct line.

# APPENDIX H

## LETTER TO PARENTS OF PRE-SCHOOL CHILDREN

12 January 1977

Dear Parent

I am writing to ask if you will co-operate in a new survey which is being carried out in Birmingham into the level of lead in the blood of young children. We shall be releasing information to the Press on Tuesday, January 18, so you may well read about it in the evening papers or see a reference to it in the television news.

As you probably know, we are all exposed to lead through the food we eat and the air we breathe. Various surveys have been carried out and the results show that the levels of lead are usually within what is regarded as safe limits. There has also been a recent survey in Birmingham which showed that blood lead levels in children between 8 and 14 were reassuringly low.

It may be, however, that younger children who tend to crawl about or play on the floor and who may suck their fingers, are more likely to pick up lead. The only way we can be sure – and we are confident you will agree that we should not leave it to chance – is by making another survey of children under five. This will achieve two things. It will give us an accurate picture of what the situation really is, and it will also make a direct check of the children who take part in the survey.

We should be grateful if you would help by letting your child take part. I must stress that we have no reason to suspect that your child is in special danger. Your name has come up simply because we are covering a broad cross-section of the population of the City.

If you agree to take part we shall arrange to take you and your child to Good Hope Hospital – we shall provide the transport if necessary – where a single drop of blood will be taken from the end of a finger. This method is chosen to ensure that the child feels the least possible pain.

In addition you will be asked to provide answers to a short questionnaire, and someone will be available to help. The questions will be very simple and will include such things as your husband's work. We need to know that in order to make sure that we do not overlook any possibility of lead getting to your home from a place of work.

You and the child will then be taken home. The results of all the tests will be analysed later and a report will be prepared both for the Government and the City of Birmingham. The report will be made public but no names will be given. The survey is entirely confidential. If you wish, you yourself will be given the result of the test and what it means. If by any chance the level of lead is thought to be high, your family doctor will be told at once and action will be taken to trace the cause and to give your child any medical treatment that may be necessary.

Please give us your help. We realise there may be some inconvenience and we and the city authorities will make it as easy for you as we possibly can. You, for your part, will have the satisfaction of knowing that you have made a contribution to an important piece of research about matters affecting the health of young children, and have had a direct check made of your own child's blood-lead level.

If you are agreeable, please fill in the slip which accompanies this letter and post it in the pre-paid envelope provided. If there are any questions you would like to ask before making up your mind, please telephone 235 4641 and ask for Dr. Nicol.

Yours faithfully
Dr. Nicol
Area Medical Officer

# APPENDIX I

## ASSESSMENT OF THE HOME ENVIRONMENT

**Method**

1. Wherever cases of enhanced exposure in pre-school children were confirmed by lead concentrations in excess of 35 $\mu$g/100 ml in the second samples of blood, an assessment of the child's home environment was carried out. Seven houses in all were examined.

2. Samples of dust were taken by vacuum and brush-cleaning methods from internal surfaces such as window frames and sills, door frames, carpets, linoleum, soft furnishings, tops of wardrobes and from clothes. Samples of paint from window sills and toys and of flaking paint were also taken where these were available.

3. Dust samples were taken by the same vacuum and brush-cleaning methods from external surfaces including window frames and sills, backyards, house eaves, pavements and road gutters. Some paint flakes were also taken. Soil in front and rear gardens was sampled.

4. Drinking water samples (300 ml) were taken by the householder in each house. In each case, 3 types of sample were collected; "first draw" water (i.e. that taken first thing in the morning before any other water was used); during the daytime, after flushing the tap for a short period of time (5 minutes); and a random spot sample taken during the daytime without any flushing.

5. A number of other factors thought potentially important were also examined. These included:
   (i) the habit of pica in the child;
   (ii) the general state of repair of paintwork, putty and plaster, with particular reference to presence of flaking paint;
   (iii) evidence of recent redecoration;
   (iv) the proximity of the home to a major road;
   (v) the growing of vegetables at home or on an allotment;
   (vi) the presence of lead plumbing in the house;
   (vii) the period of residence at the home and the previous address;
  (viii) the presence of lead in cooking utensils, cosmetics, medicines, toys and household objects, and the use of any special culinary powders;
   (ix) the possible exposure to lead from hobbies e.g. from glazed vessels used in wine-making;
   (x) a further check on parents' occupations;

(xi) the proximity of works using lead or lead compounds to the home;
(xii) the proximity of works using lead or lead compounds to parents' places of employment.

6. The extent and depth of the survey was limited by difficulties of communication and cooperation with some of the children's parents, and in some cases it was not possible to obtain all the required information.

## Results

7. The parents' occupations showed no evidence of exposure to lead at work; dust from clothing showed no evidence of transport of lead into the home from the workplace.

8. No evidence was found that lead in cooking utensils, cosmetics, medicines, toys and household objects, or special culinary powders, played any significant role in enhanced exposure.

9. All the families had lived in their present houses for at least a year and up to 14 years in one case. This indicates that the child's blood lead concentration, reflecting recent exposure, is unlikely to be enhanced due to previous exposure associated with a different environment. In two of the three cases, where a previous address was given, the change of address occurred within areas of similar type, which would not therefore be expected to have resulted in a marked change in exposure to airborne lead; in the third case, the move was such as to indicate a decrease in exposure to airborne lead.

10. Garden vegetables were grown by only two families. Although soil samples in these cases had lead concentrations of 900 and 990 ppm, uptake of lead by the roots of plants and translocation to the edible parts of the plants occurs only to a very limited extent and it is unlikely that such soil concentrations could result in significantly enhanced total dietary exposure. Soil samples from gardens where vegetables were not grown ranged from 260–1,850 ppm; in two cases, the higher value was seen in the front garden, perhaps because this was nearer the road.

11. There was no evidence that the homes were affected to a significant extent by the proximity of works using lead or lead compounds. At one house 750 metres from a large battery factory, a high concentration of lead (900 ppm) was found in a dust sample taken from the eaves' gutter but other external samples at this house were within the normal range for urban areas (range of four samples 820–1,960 ppm). The proximity of works using lead or lead compounds to the parents' places of employment did not seem, on the basis of lead concentrations in parents' clothing, to represent a significant source of exposure.

12. Four of the houses were near major roads, in two cases close to the Stratford Road. Monitoring has been carried out near the Stratford Road and the results discussed in Chapter 2. In view of the fact that airborne lead concentrations are not exceptional in such situations, it seems reasonable to conclude that respiration of airborne lead from traffic is unlikely to be a significant cause of undue exposure in these two cases. Similarly, there is no reason to suppose that Coventry Road and Little Sutton Road are significant causes of enhanced exposure through respiration.

13. Concentration of lead in drinking water ranged overall from <0·01–0·21 mg/l. In the two houses in Sutton Coldfield, concentrations in excess of 0·01 mg/l were not seen: for comparison, the maximum lead concentration recommended by the World Health Organisation (WHO) is currently 0·1 mg/l; drinking water seems not to be responsible for enhanced exposure in these cases. In the other five houses in the central areas, concentrations above 0·1 mg/l were seen on only three occasions and then only in "first draw" water, for which the WHO recommends a limit of 0·3 mg/l: the remaining samples (21) showed a range of lead concentrations of <0·01–0·09 mg/l, with a mean of 0·026 mg/l. Again it seems very unlikely, in the light of recent work on the relationship between blood lead and water lead concentrations (Moore et al, 1977) that drinking water could be a cause of undue exposure in these children.

14. As is clear from the above discussion, most of the factors examined appeared not to be potential causes of enhanced exposure. This is a common finding in work of this kind. However, it has previously been reported (Baker et al, 1977) that dust lead concentrations show significant correlations with raised blood lead concentrations and the results for dust from the present survey are discussed below.

15. In one of the houses in the central area, lead concentrations in internal dust samples ranged 750–2,900 ppm, with a mean of 1,404 ppm; external dusts ranged 520–1,200 ppm lead (mean 874 ppm). These values are quite typical of urban conditions. The soil lead concentration was 700 ppm and whilst this is above the normal range for unpolluted soils it is not unusual for urban conditions. None of the Asian children living in this house exhibited pica and it is impossible to attribute their raised blood lead concentrations to any particular source.

16. In another of the houses in the central area, dust lead concentrations in indoor samples ranged 610–3,590 ppm (five samples) with one sample at 8,950 ppm. The latter was taken from a room used solely as a store. As noted in paragraph 11, concentrations in external dusts ranged 820–1,960 ppm (four samples) with one sample at 22,900 ppm; this was taken from roof guttering. Soil concentrations were 340 ppm (back garden) and 1,850 ppm (front garden). Whilst the mother was uncertain whether the child exhibited pica, it

is difficult to attribute enhanced exposure to dust, given the inaccessible nature of the only two samples with raised lead concentrations and the finding of Barltrop et al (1975) that substantially raised blood lead levels are only seen, even in children with pica, when soil lead concentrations exceed about 10,000 ppm.

17.   The third house in the central area showed internal dust lead concentrations of 520–11,600 ppm, with paint flakes having a concentration of 21,100 ppm lead. External dust samples showed a range of 560–7,970 ppm lead (the latter on painted bricks) with paint flakes on an external toilet door having 24,300 ppm lead. Soil lead concentrations were 650 ppm (back) and 1,080 ppm (front). Again it is difficult to attribute undue exposure of the Asian child to any particular cause, particularly because there was no evidence of pica, but the flaking paint seems a possible cause.

18.   Two Asian children at a fourth house in the central area exhibited pica. Nevertheless, apart from two samples showing lead concentrations of 9,500 ppm and 28,000 ppm, internal dusts had lead concentrations in the range 820–4,400 ppm (mean of 16 samples – 2,151 ppm); and external dusts and soils were in the range 850–3,300 ppm lead (mean 1,687 ppm). Clearly it is possible that isolated lead-rich samples, if ingested by children, could result in temporarily raised blood lead concentrations; but it seems unlikely that a confirmed and longer-lasting raised exposure could be due to such isolated dust samples.

19.   A somewhat similar situation occurred at the fifth house in the central area. Here, one internal dust sample had a lead concentration of 29,000 ppm and one had 8,900 ppm, whilst the mean of nine other samples was 1,208 ppm (range 690–1,900 ppm). Paint flakes on an external toilet door had 61,000 ppm lead, but external dusts were in the range 550–1,100 ppm lead. The child was known to pick paint flakes from the toilet door, but the mother was uncertain whether he chewed them. It seems possible that this was the source of undue exposure.

20.   Two houses in Sutton Coldfield were examined. At the first, internal dust samples ranged 490–9,560 ppm. The higher levels were seen in rooms facing a busy road and one of these was the child's bedroom. External dust samples ranged 180–4,600 ppm and the soil lead concentration was 260 ppm. The child was known to exhibit pica for materials found both inside and outside the home and it is possible here that house dusts were the cause of raised blood lead concentrations. However, a concentration approaching 10,000 ppm in dusts seems unlikely to be the result solely of traffic-generated dusts and it may be significant that redecoration had recently taken place, perhaps resulting in deposition of lead-rich dusts from the sanding of old lead-containing paints.

21. The clearest example of a heavily contaminated environment occurred in an old house in Sutton Coldfield where extensive redecoration had recently been carried out. Internal dust samples in the typical range for urban conditions were restricted to five samples with concentrations of 2,040–4,470 ppm lead; by contrast seven dust samples had lead concentrations of 6,040–78,800 ppm and paint flakes were found with a concentration of 162,400 ppm; the mean in these eight was 49,965 ppm. The father's jacket, worn during the redecorating, contained dust with a lead concentration of 10,600 ppm. External dusts and garden soil had lead concentrations in the range 660–1,300 ppm. The children were known to exhibit symptoms of pica and it seems reasonable to conclude that ingestion of paint flakes or dusts containing lead derived from sanding old lead-containing paints or inhalation of such dusts, was the cause of enhanced exposure.

**Conclusions**
22. Most of the factors thought to be potential causes of enhanced lead exposure were not implicated by the environmental assessment. These included parental occupation, consumption of home-grown vegetables, respiration of airborne lead from traffic, and lead in drinking water. There was no evidence of the use of lead-containing cooking utensils, cosmetics, medicines, or special culinary powders. Similarly, there was no evidence that nearness of the home or the parents' places of work to any industrial source of lead played any significant role in enhanced exposure.

23. In one household it was impossible to attribute exposure to any particular source of lead. In two other houses, flaking paint with high lead concentrations existed but the absence of or uncertainty about pica makes definition of a specific source difficult. At three other houses, isolated or inaccessible samples of lead-rich dust were found but it is difficult to be sure that these were the cause of enhanced exposure, even though pica was known to occur in two of the households.

24. In only one house was the cause of undue exposure found with a reasonable degree of certainty. Extensive contamination by lead-rich dust containing lead derived from the preparation of old lead paint existed and extensive redecoration was consistent with this conclusion. Pica was noted in the exposed children. It seems likely that this could be a general problem where preparation of old paint surfaces is carried out by sanding and where children are known to exhibit pica. It may be preferable for old paint to be removed by chemical stripping – which avoids production of dust – rather than by sanding or burning, both of which produce particulate matter.

Figure 1: An aerial view of the interchange looking
south-east

49

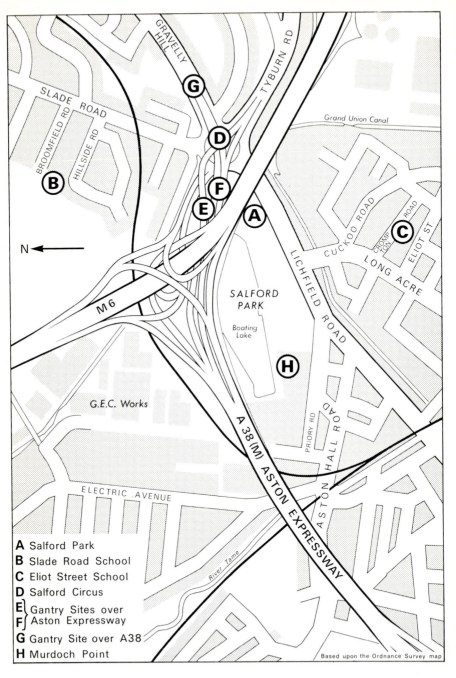

A Salford Park
B Slade Road School
C Eliot Street School
D Salford Circus
E⎫ Gantry Sites over
F⎭ Aston Expressway
G Gantry Site over A38
H Murdoch Point

Based upon the Ordnance Survey map

Figure 2: Location of lead monitoring sites

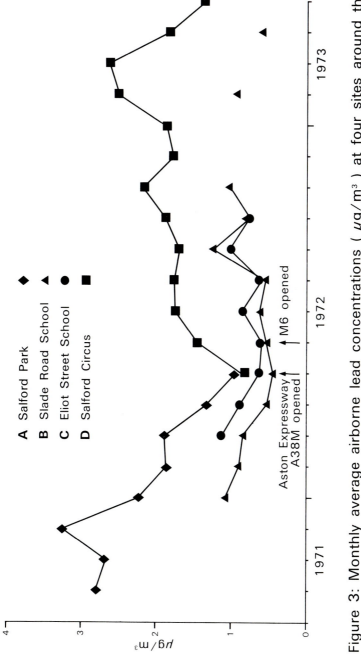

Figure 3: Monthly average airborne lead concentrations ( $\mu g/m^3$ ) at four sites around the M6–A38M interchange

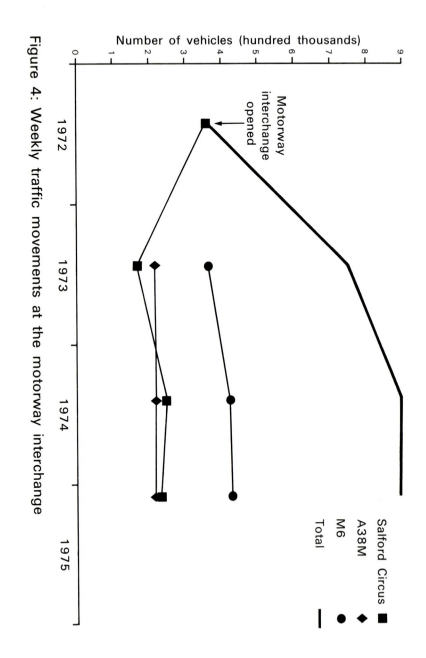

Figure 4: Weekly traffic movements at the motorway interchange

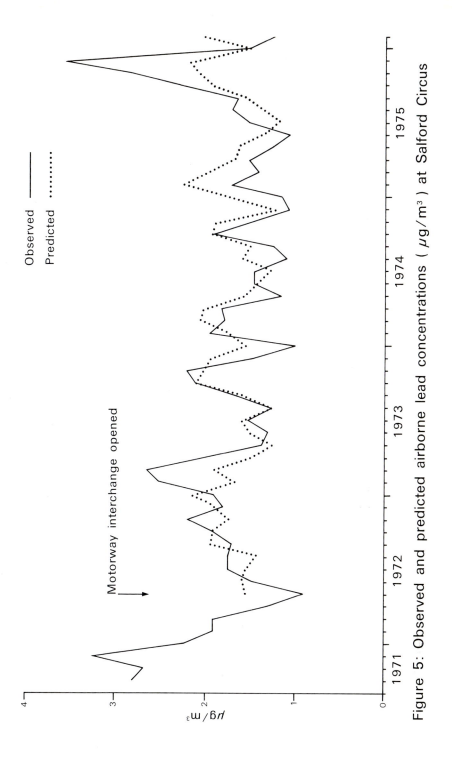

Figure 5: Observed and predicted airborne lead concentrations ($\mu$g/m³) at Salford Circus

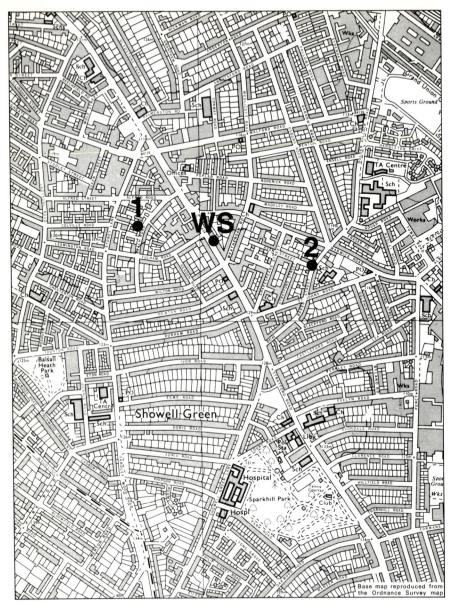

Figure 6: Location of urban monitoring sites

Site 1 at Longlands House 270 metres west of kerbside monitoring site (**WS**) beside the A34

Site 2 at Arden School 335 metres east of monitoring site (**WS**)

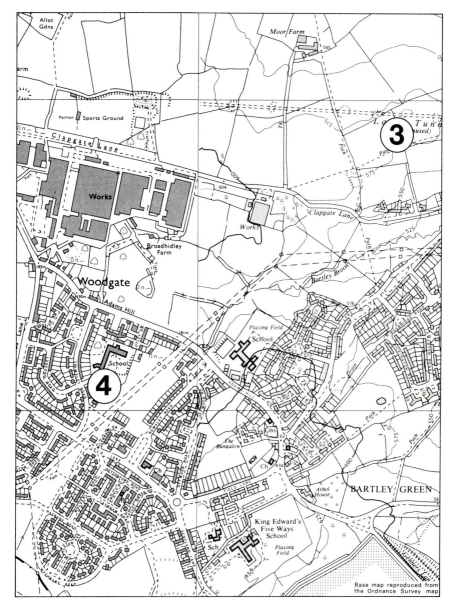

Figure 7: Location of suburban monitoring sites

Site **3**   Woodgate Valley Association Hall

Site **4**   Milebrook Road Community Centre

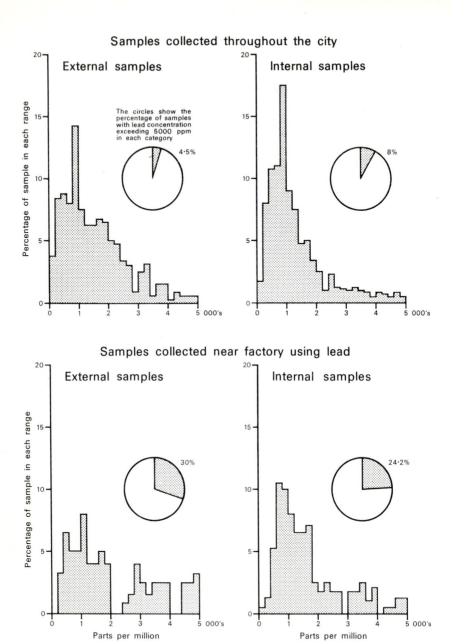

Figure 8: Frequency distributions* of lead concentrations
in dust samples

*Reproduced by permission of the Royal Society of Health Journal.

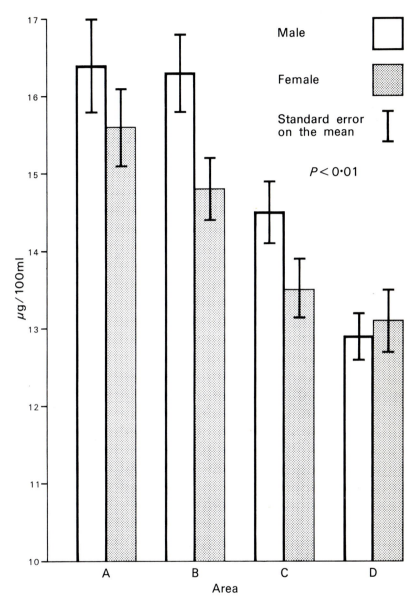

Figure 10: Trends in blood lead concentration
( $\mu$g/100ml ) in schoolchildren by area

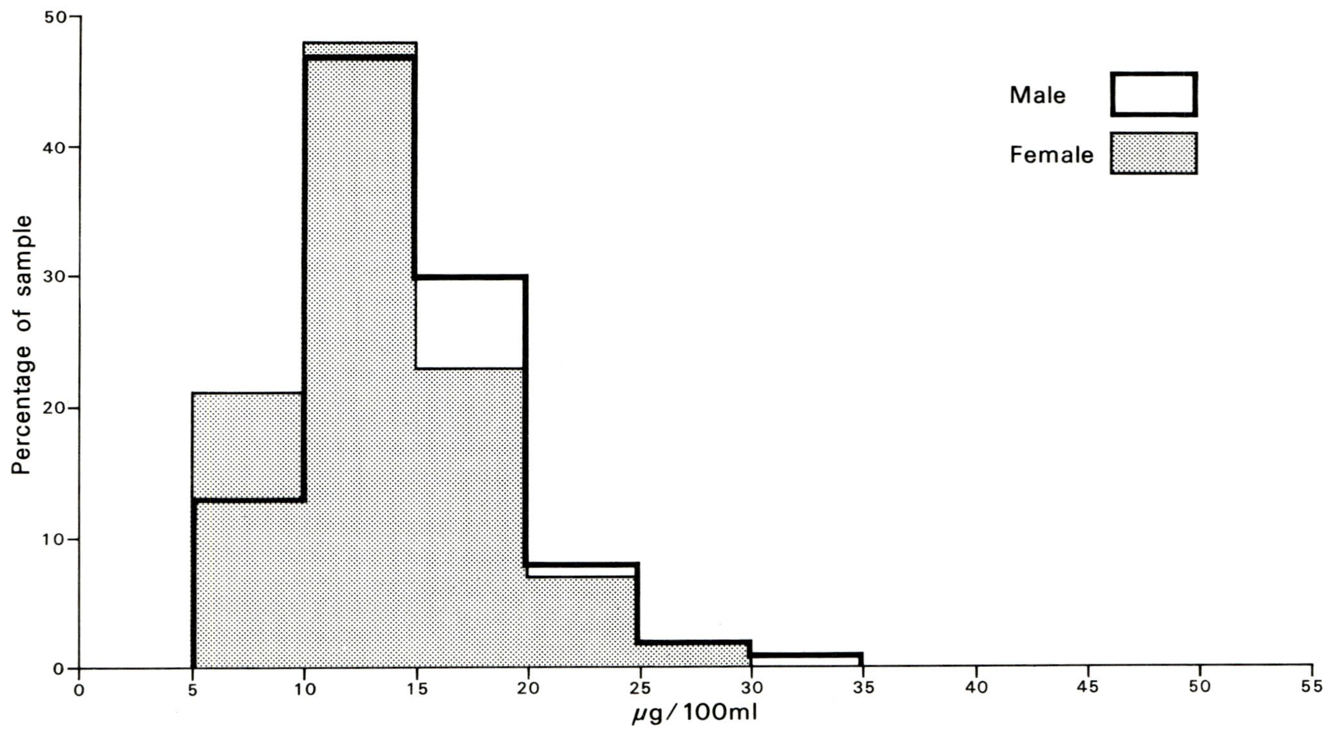

Figure 11: Frequency distribution of blood lead concentration ( $\mu$g/100ml ) in schoolchildren

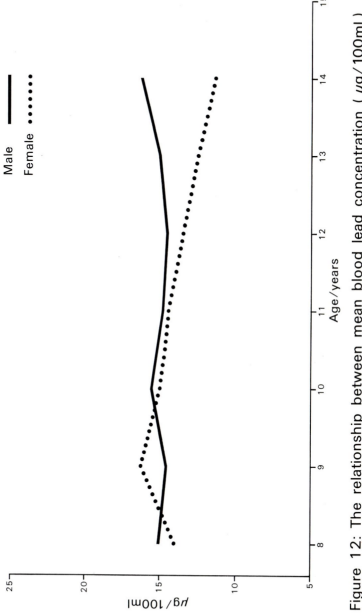

Figure 12: The relationship between mean blood lead concentration ( $\mu$g/100ml ) and age : schoolchildren

61

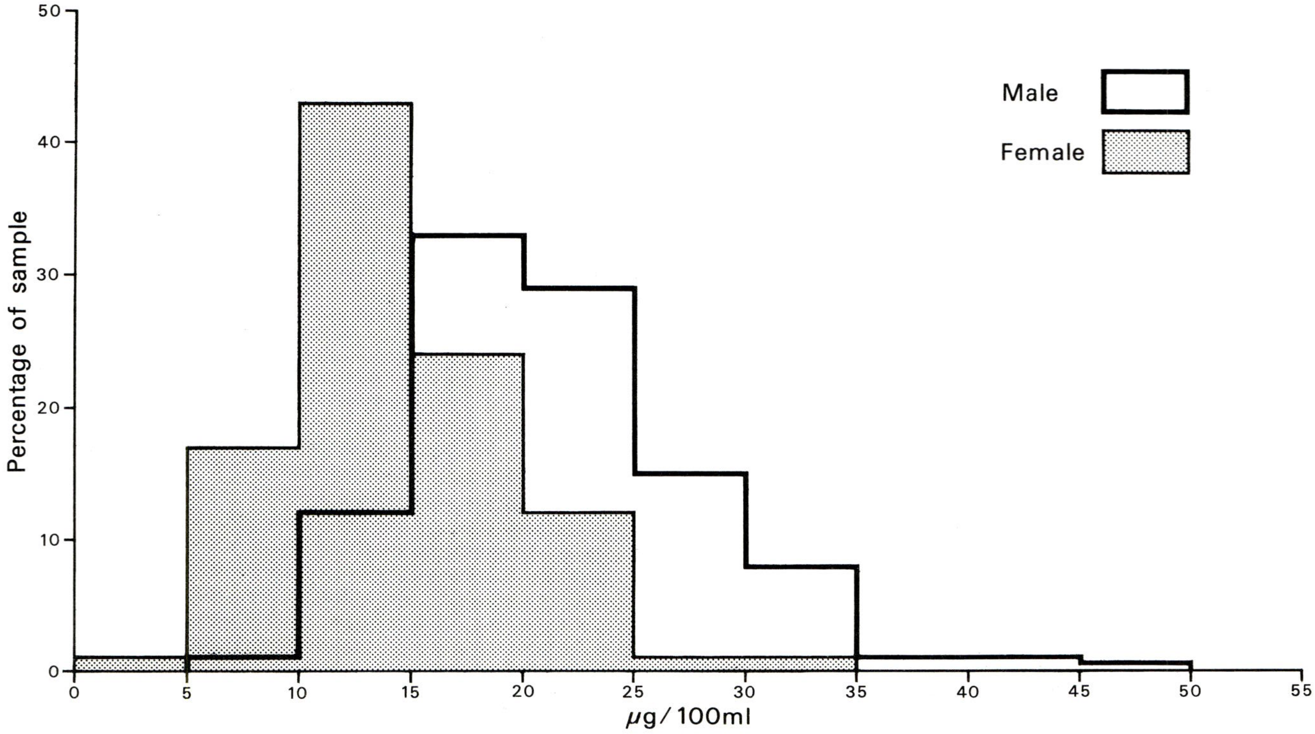

Figure 13: Frequency distribution of blood lead concentration ( $\mu g$ / 100ml ) in adults

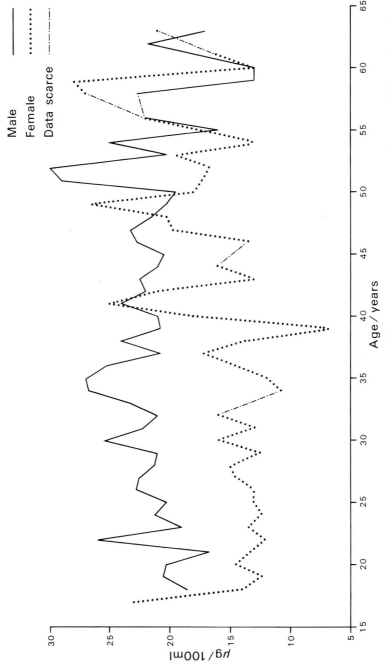

Figure 14: The relationship between mean blood lead concentration ( $\mu g/100ml$ ) and age : adults

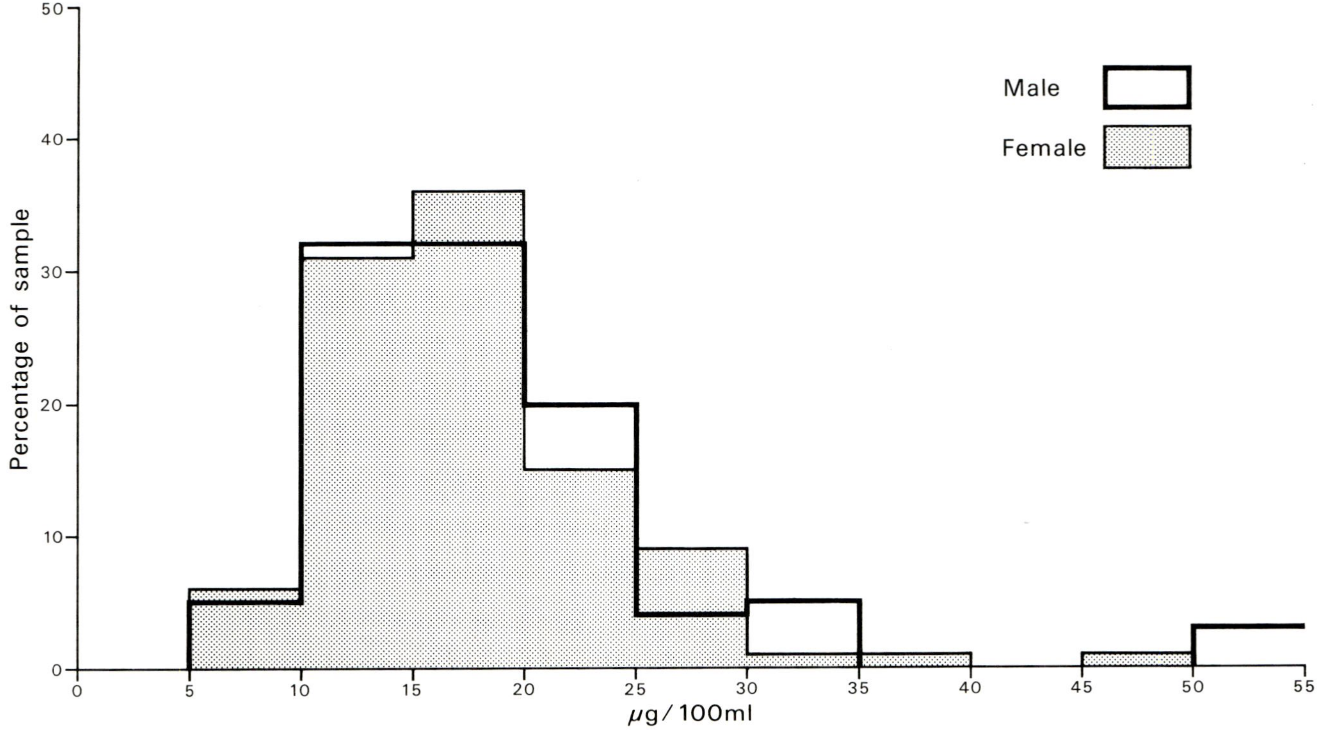

Figure 15: Frequency distribution of blood lead concentration ( $\mu$g/100ml ) in randomly selected pre-school children

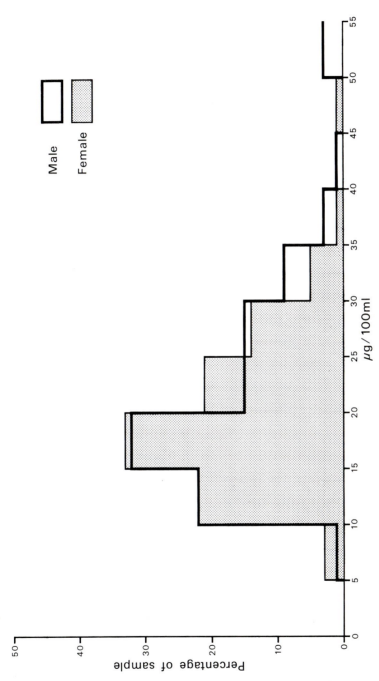

Figure 16: Frequency distribution of blood lead concentration ( $\mu$g/100ml ) in non-randomly selected pre-school children

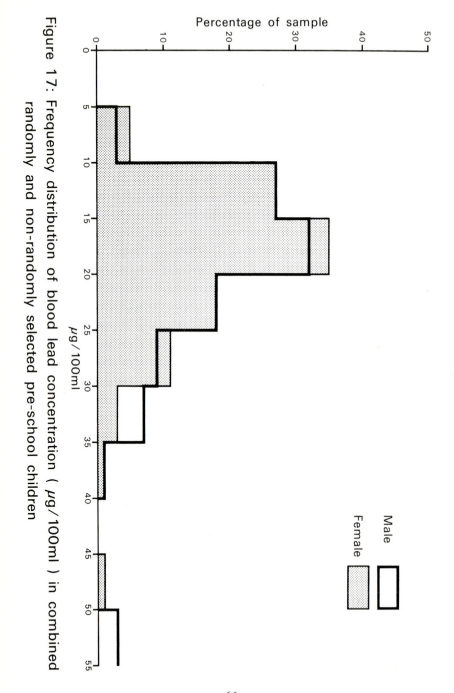

Figure 17: Frequency distribution of blood lead concentration ( $\mu g$ / 100ml ) in combined randomly and non-randomly selected pre-school children

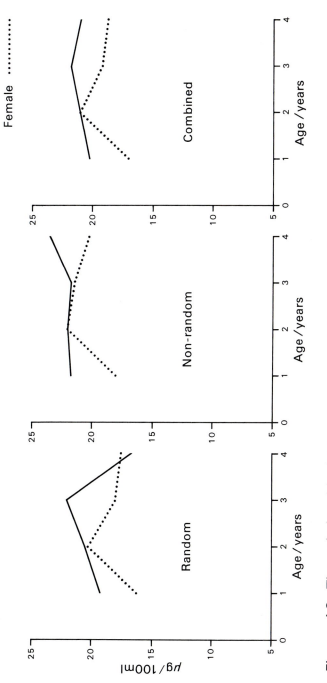

Figure 18: The relationship between mean blood lead concentration ( μg/100ml ) and age : sample groups

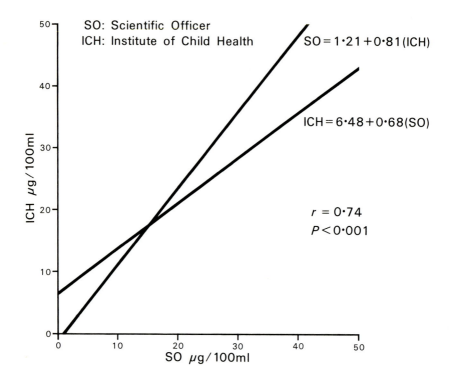

Figure 19: Regression between analytical laboratories of blood lead concentrations ( μg/100ml ) in venous samples from schoolchildren

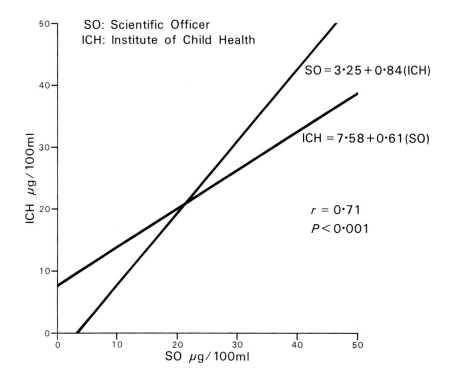

Figure 20: Regression between analytical laboratories of blood lead concentrations ( μg/100ml ) in capillary samples from pre-school children

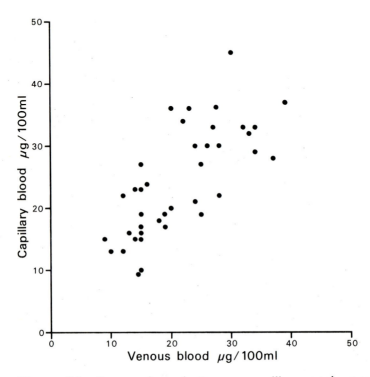

Figure 21: Comparison between capillary and venous blood lead concentrations ( $\mu$g/100ml ) in adults

# TABLE 1

Monthly average lead concentrations, 'day' samples
($\mu g/m^3$)

| Month | A | B | C | D | E | F | G | H |
|---|---|---|---|---|---|---|---|---|
| | Salford Park | Slade Rd School | Eliot St School | Salford Circus | Gantry (Bottom) Expressway | Gantry (Top) Expressway | Gantry Gravelly Hill | Murdoch Point (Top) |
| Oct. 71 | 3·51 | | | | | | | |
| Nov. | 3·22 | | | | | | | |
| Dec. | 3·85 | | | | | | | |
| Jan. 72 | 3·01 | 1·16 | | | | | | |
| Feb. | 2·47 | 1·03 | | | | | | |
| Mar. | 2·01 | 0·90 | 0·99 | | | | | |
| Apr. | 1·59 | 0·51 | 0·93 | | | | | |
| May | 1·12 | 0·50 | 0·78 | 1·03 | | | | |
| June | | 0·51 | 0·64 | 1·70 | | | | |
| July | | 0·60 | 0·66 | 2·03 | | | | |
| Aug. | | 0·51 | 0·54 | 2·29 | | | | |
| Sept. | | 1·45 | 1·02 | 2·27 | | | | |
| Oct. | | 0·81 | 0·76 | 2·16 | | | | |
| Nov. | | 1·25 | | 2·97 | | | | |
| Dec. | | | | 2·29 | 2·24 | 2·39 | | |
| Jan. 73 | | | | 2·59 | 2·76 | 2·21 | | |
| Feb. | | 1·12 | | 3·36 | | 1·99 | | |
| Mar. | | | | 2·69 | | | | |
| Apr. | | 0·52 | | 2·15 | | | 0·73 | |
| May | | | | 1·65 | | | | 0·38 |

# TABLE 2

Monthly average lead concentrations, 'night' samples
($\mu g/m^3$)

| Month | A | B | C | D | E | F | G | H |
|---|---|---|---|---|---|---|---|---|
| | Salford Park | Slade Rd School | Eliot St School | Salford Circus | Gantry (Bottom) Expressway | Gantry (Top) Expressway | Gantry Gravelly Hill | Murdoch Point (Top) |
| Oct. 71 | 2·10 | | | | | | | |
| Nov. | 2·14 | | | | | | | |
| Dec. | 2·68 | | | | | | | |
| Jan. 72 | 1·46 | 1·00 | | | | | | |
| Feb. | 1·32 | 0·81 | | | | | | |
| Mar. | 1·78 | 0·83 | 1·24 | | | | | |
| Apr. | 1·08 | 0·53 | 0·83 | | | | | |
| May | 0·67 | 0·43 | 0·50 | 0·69 | | | | |
| June | | 0·55 | 0·61 | 1·20 | | | | |
| July | | 0·68 | 1·05 | 1·45 | | | | |
| Aug. | | 0·56 | 0·60 | 1·24 | | | | |
| Sept. | | 1·05 | 1·07 | 1·12 | | | | |
| Oct. | | 0·72 | 0·71 | 1·58 | | | | |
| Nov. | | 0·78 | | 1·41 | | | | |
| Dec. | | | | 1·28 | 1·08 | 0·91 | | |
| Jan. 73 | | | | 1·19 | 1·41 | 0·74 | | |
| Feb. | | 0·74 | | 1·63 | | 1·29 | | |
| Mar. | | | | 2·58 | | | | |
| Apr. | | 0·68 | | 1·92 | | | 0·76 | |
| May | | | | 1·09 | | | | 0·51 |

**TABLE 3**

Combined day and night monthly average lead concentrations
($\mu$g/m$^3$)

| Month | A | B | C | D | E | F | G | H |
|---|---|---|---|---|---|---|---|---|
| | Salford Park | Slade Rd School | Eliot St School | Salford Circus | Gantry (Bottom) Express-way | Gantry (Top) Express-way | Gantry Gravelly Hill | Murdoch Point (Top) |
| Oct. 71 | 2·80 | | | | | | | |
| Nov. | 2·68 | | | | | | | |
| Dec. | 3·26 | | | | | | | |
| Jan. 72 | 2·23 | 1·08 | | | | | | |
| Feb. | 1·89 | 0·92 | | | | | | |
| Mar. | 1·89 | 0·86 | 1·11 | | | | | |
| Apr. | 1·33 | 0·52 | 0·88 | | | | | |
| May | 0·97 | 0·46 | 0·64 | 0·86 | | | | |
| June | | 0·53 | 0·62 | 1·45 | | | | |
| July | | 0·64 | 0·85 | 1·74 | | | | |
| Aug. | | 0·53 | 0·57 | 1·76 | | | | |
| Sept. | | 1·25 | 1·04 | 1·69 | | | | |
| Oct. | | 0·76 | 0·73 | 1·87 | | | | |
| Nov. | | 1·01 | | 2·19 | | | | |
| Dec. | | | | 1·77 | 1·66 | 1·65 | | |
| Jan. 73 | | | | 1·89 | 2·08 | 1·47 | | |
| Feb. | | 0·93 | | 2·50 | | 1·64 | | |
| Mar. | | | | 2·63 | | | | |
| Apr. | | 0·60 | | 1·84 | | | 0·75 | |
| May | | | | 1·57 | | | | 0·44 |

72

**TABLE 4**

Maximum 12-hour lead concentrations recorded in each month
($\mu g/m^3$)

| Month | Salford Park | | Slade Road | | Eliot Street | | Salford Circus | | Gantry (Base) Expressway | | Gantry (Top) Expressway | | Gantry Gravelly Hill | | Murdoch Point | |
|---|---|---|---|---|---|---|---|---|---|---|---|---|---|---|---|---|
| | Day | Night | Day | Night | Day | Night | Day | Night | Day | Night | Day | Night | Day | Night | Day | Night |
| Oct. 71 | 8·24 | 4·48 | | | | | | | | | | | | | | |
| Nov. | 6·35 | 12·22 | | | | | | | | | | | | | | |
| Dec. | 12·14 | 11·25 | | | | | | | | | | | | | | |
| Jan. 72 | 10·55 | 3·70 | 2·43 | 2·87 | | | | | | | | | | | | |
| Feb. | 4·65 | 2·22 | 2·07 | 1·73 | | | | | | | | | | | | |
| Mar. | 4·26 | 4·37 | 2·10 | 2·26 | 2·73 | 5·57 | 1·83 | 1·59 | | | | | | | | |
| Apr. | 3·05 | 2·35 | 1·44 | 1·70 | 4·64 | 4·24 | 2·73 | 2·95 | | | | | | | | |
| May | 3·05 | 2·15 | 2·46 | 1·14 | 2·48 | 1·47 | 4·49 | 4·72 | | | | | | | | |
| June | | | 1·31 | 1·56 | 1·02 | 1·69 | 9·34 | 2·73 | | | | | | | | |
| July | | | 1·93 | 2·62 | 1·85 | 7·94 | 9·07 | 4·55 | | | | | | | | |
| Aug. | | | 1·15 | 1·70 | 1·30 | 1·35 | 7·91 | 3·25 | | | | | | | | |
| Sept. | | | 1·54 | 1·48 | 0·78 | 0·79 | 8·13 | 4·28 | | | | | | | | |
| Oct. | | | 1·10 | 0·95 | | | 7·40 | 3·66 | | | | | | | | |
| Nov. | | | 2·10 | 1·98 | | | 6·53 | 2·59 | | | | | | | | |
| Dec. | | | | | | | 9·59 | 4·20 | 4·00 | 3·62 | 7·78 | 2·65 | | | | |
| Jan. 73 | | | | | | | 5·68 | 7·71 | | | | | | | | |
| Feb. | | | 6·60 | 1·80 | | | | | | | | | | | | |
| Mar. | | | | | | | | | | | | | | | | |
| Apr. | | | | | | | | | | | | | 1·36 | 1·23 | | |
| May | | | | | | | | | | | | | | | 0·67 | 0·99 |

73

## TABLE 5

Monthly average values of percentage lead of the total particulate collected during early sampling period

| Month | Salford Park | | | Slade Road School | | | Eliot Street School | | | Salford Circus | | |
|---|---|---|---|---|---|---|---|---|---|---|---|---|
| | Day | Night | Combined average | Day | Night | Combined average | Day | Night | Combined average | Day | Night | Combined average |
| October | 1·99 | 1·83 | 1·91 | | | | | | | | | |
| November | 1·76 | 1·69 | 1·72 | | | | | | | | | |
| December | 3·06 | 3·22 | 3·14 | | | | | | | | | |
| 1972 | | | | | | | | | | | | |
| January | 1·03 | 0·74 | 0·88 | 1·08 | 0·81 | 0·94 | | | | | | |
| February | 1·31 | 0·77 | 1·04 | 1·18 | 0·80 | 0·99 | | | | | | |
| March | 1·28 | 0·89 | 1·08 | 0·75 | 0·71 | 0·73 | 0·52 | 0·77 | 0·64 | | | |
| April | 0·99 | 1·30 | 1·14 | 0·47 | 0·95 | 0·71 | 0·61 | 0·86 | 0·74 | | | |
| May | 0·61 | 0·44 | 0·52 | 0·56 | 0·55 | 0·56 | 0·77 | 0·62 | 0·69 | 0·57 | 0·63 | 0·60 |
| June | | | | 0·44 | 0·45 | 0·45 | 0·48 | 0·65 | 0·56 | 1·73 | 1·49 | 1·61 |
| July | | | | 0·61 | 0·73 | 0·67 | 0·81 | 1·01 | 0·91 | 1·40 | 1·43 | 1·41 |
| August | | | | | | | | | | 1·52 | 1·38 | 1·45 |
| | Mean 1·43 | | | Mean 0·72 | | | Mean 0·71 | | | Mean 1·27 | | |

## TABLE 6

Airborne lead concentrations at Salford Circus
($\mu$g/m$^3$)

| | 1972 | 1973 | 1974 | 1975 | 1976 | Mean | Standard deviation |
|---|---|---|---|---|---|---|---|
| Jan. | | 1·90 | 0·99 | 1·17 | 1·48 | 1·38 | 0·37 |
| Feb. | | 2·50 | 1·95 | 1·70 | 1·17 | 1·83 | 0·48 |
| Mar. | | 2·63 | 1·79 | 1·38 | | 1·93 | 0·53 |
| Apr. | | 2·04 | 1·83 | 1·50 | | 1·79 | 0·22 |
| May | 0·86 | 1·37 | 1·15 | 1·25 | | 1·16 | 0·17 |
| June | 1·45 | 1·30 | 1·47 | 1·02 | | 1·31 | 0·17 |
| July | 1·74 | 1·53 | 1·48 | 1·50 | | 1·56 | 0·14 |
| Aug. | 1·76 | 1·31 | 1·10 | 1·70 | | 1·47 | 0·26 |
| Sept. | 1·69 | 1·66 | 1·25 | 1·62 | | 1·56 | 0·14 |
| Oct. | 1·87 | 2·11 | 1·94 | 2·23 | | 2·04 | 0·19 |
| Nov. | 2·19 | 2·19 | 1·55 | 2·86 | | 2·20 | 0·45 |
| Dec. | 1·77 | 1·42 | 1·07 | 3·56 | | 1·96 | 0·95 |
| Mean | 1·67 | 1·83 | 1·46 | 1·79 | | | |
| Standard deviation | 0·34 | 0·20 | 0·35 | 0·72 | | | |

## TABLE 7

| Airborne lead concentrations ($\mu$g/m$^3$) in urban and suburban sites | | | | |
|---|---|---|---|---|
| | Urban[a] | | Suburban[b] | |
| | 1 | 2 | 3 | 4 |
| Day-time mean | 1·17 | 1·18 | 0·60 | 0·60 |
| Night-time mean | 1·00 | 1·16 | 0·64 | 0·52 |
| Combined mean | 1·08 | 1·17 | 0·62 | 0·56 |

1. Longland House
2. Arden School
3. Woodgate Valley Tenants Association Hall
4. Milebrook Community Centre

[a]Sampling periods 6 November 1975 – 11 December 1975.
[b]Sampling periods 18 February 1976 – 16 March 1976.

**TABLE 8**

Atmospheric lead concentrations ($\mu$g/m$^3$) inside and outside houses

1. House 160 m from Salford Circus monitored between 29 November and 19 December 1973.

|  | Inside | Outside | Salford Circus | Percentage lead inside/outside |
|---|---|---|---|---|
| Day-time | 1·07 | 1·14 | 2·16 | 94 |
| Night-time | 0·66 | 0·62 | 0·98 | 106 |
| Combined mean | 0·87 | 0·88 | 1·57 | 98 |

2. House alongside the M6 at Perry Barr, Birmingham monitored between 1 March and 22 March 1974

|  | | | | |
|---|---|---|---|---|
| Day-time | 0·70 | 1·01 | 2·99 | 69 |
| Night-time | 0·68 | 0·96 | 1·53 | 71 |
| Combined mean | 0·69 | 0·98 | 2·26 | 70 |

3. House with double-glazing alongside the M6 at Perry Barr, Birmingham monitored between 26 April and 15 May 1974

|  | | | | |
|---|---|---|---|---|
| Day-time | 0·49 | 0·75 | 1·56 | 65 |
| Night-time | 0·39 | 0·66 | 0·97 | 59 |
| Combined mean | 0·44 | 0·70 | 1·26 | 63 |

**TABLE 9**

Lead concentrations in dust samples

| Sample classification | Median lead concentration ppm by weight | | Number of samples | | Percentage of samples with more than 5,000 ppm lead, by weight | |
|---|---|---|---|---|---|---|
| | A | B | A | B | A | B |
| All samples from roads | 1,700 | 1,200 | 234 | 1,061 | 9·9 | 2·7 |
| External samples other than from roadside | 3,000 | 1,200 | 127 | 455 | 30·0 | 4·5 |
| All internal samples | 1,600 | 1,000 | 240 | 806 | 24·2 | 8·0 |

A = samples obtained within a 400 m radius of a lead battery factory.
B = samples obtained throughout the City, but excluding those from area A.

77

## TABLE 10

Lead concentrations in dust samples from different sites[a]

| Category | Sample classification | Sample population | Median lead concentration (ppm) | Percentage of samples in class > 5,000 ppm |
|---|---|---|---|---|
| 1[b] | All samples from roads | 1,061 | 1,200 | 2·7 |
| 2 | Roadside gutter in major arterial or ring roads ('A' class) | 151 | 1,800 | 5·5 |
| 3 | Roadside gutter in heavily trafficked subsidiary road ('B' class) | 92 | 1,300 | 0 |
| 4 | Roadside gutter in residential road | 89 | 950 | 0 |
| 5 | Pavement adjacent to residential road | 181 | 700 | 0 |
| 6 | Roadside gutter in industrial estate road | 115 | 1,300 | 6·2 |
| 7 | Roadside gutter in mixed residential and industrial road | 164 | 950 | 3·2 |
| 8 | All internal samples | 806 | 1,000 | 8·0 |
| 9 | Samples from residential floor coverings only | 538 | 950 | 6·5 |
| 10 | All external surfaces excluding roadside samples | 455 | 1,200 | 4·5 |

[a]Excluding the area around the battery factory.
[b]Category 1 includes categories 2–7 inclusive, together with samples from other miscellaneous sources.

**TABLE 11**

Relative amounts of lead accumulated on socks

| Group | 1 | 2 | 3 | 4 | 5 | 6 | 7 |
|---|---|---|---|---|---|---|---|
| Number in group | 5 | 3 | 19 | 3 | 8 | 3 | 1 |
| Mean lead value (mg/sock) | 7·3 | 17·4 | 33·9 | 39·8 | 0·16 | 0·05 | 12·9 |
| Range (mg/sock) | 0·36–26·7 | 4·2–28·5 | 4·3–102·0 | 5·4–73·4 | 0·09–0·36 | 0·03–0·06 | |

| Group | Function |
|---|---|
| 1 | Administration |
| 2 | Supervisory |
| 3 | Manual – inside |
| 4 | Manual – outside |
| 5 | Control group |
| 6 | Blank samples |
| 7 | Others |

The men in groups 1–4 were all employees of the lead battery factory. The individual in group 7 visited the battery factory and also spent several hours sanding paintwork at his home.

# TABLE 12

Number of school children for whom permission was given to participate in the blood lead study

| Age | Area A | | B | | C | | D | | Totals | |
|---|---|---|---|---|---|---|---|---|---|---|
| | Boys | Girls | Boys | Girls | Boys | Girls | Boys | Girls | Boys | Girls |
| 8 | 5 | 9 | 9 | 17 | 20 | 18 | 20 | 20 | 54 | 64 |
| 9 | 14 | 20 | 8 | 20 | 20 | 20 | 12 | 8 | 54 | 68 |
| 10 | 20 | 18 | 17 | 9 | 20 | 20 | 13 | 8 | 70 | 55 |
| 11 | 20 | 17 | 20 | 20 | 20 | 13 | 14 | 12 | 74 | 62 |
| 12 | 11 | 11 | 20 | 20 | 15 | 12 | 15 | 11 | 61 | 54 |
| 13 | 10 | 10 | 7 | 7 | 2 | 9 | 9 | 15 | 28 | 41 |
| 14 | 0 | 0 | 14 | 18 | 2 | 5 | 3 | 3 | 19 | 26 |
| | 80 | 85 | 95 | 111 | 99 | 97 | 86 | 77 | 360 | 370 |

Note that the children were classified into areas according to their home addresses, not that of their schools.

80

## TABLE 13

Blood lead concentrations ($\mu$g/100 ml) in Birmingham school children, by area

| | Area A | B | C | D | Total |
|---|---|---|---|---|---|
| *Males* | | | | | |
| Mean | 16·4 | 16·3 | 14·5 | 12·9 | 15·0 |
| Standard deviation | 4·9 | 4·9 | 3·8 | 3·1 | 4·4 |
| Standard error | 0·6 | 0·5 | 0·4 | 0·3 | 0·2 |
| Range | 7–30 | 8–32 | 7–24 | 7–22 | 7–32 |
| Number of samples | 80 | 94 | 99 | 86 | 359 |
| *Females* | | | | | |
| Mean | 15·6 | 14·8 | 13·5 | 13·1 | 14·2 |
| Standard deviation | 4·9 | 4·4 | 3·8 | 3·7 | 4·3 |
| Standard error | 0·5 | 0·4 | 0·4 | 0·4 | 0·2 |
| Range | 7–34 | 8–28 | 6–26 | 7–28 | 6–34 |
| Number of samples | 87 | 110 | 99 | 77 | 373 |

## TABLE 14

Blood lead concentrations ($\mu$g/100 ml) in Birmingham school children, by age

|  | 8 | 9 | 10 | 11 | 12 | 13 | 14 |
|---|---|---|---|---|---|---|---|
| *Males* | | | | | | | |
| Mean | 14·9 | 14·8 | 15·6 | 14·8 | 14·5 | 15·0 | 16·2 |
| Standard deviation | 4·1 | 4·6 | 5·2 | 3·8 | 4·0 | 4·8 | 4·8 |
| Standard error | 0·6 | 0·6 | 0·6 | 0·4 | 0·5 | 0·9 | 1·1 |
| Range | 7–27 | 8–30 | 7–32 | 7–24 | 7–22 | 8–28 | 10–31 |
| Number of samples | 55 | 53 | 69 | 75 | 59 | 29 | 18 |
| *Females* | | | | | | | |
| Mean | 14·4 | 16·3 | 14·9 | 14·4 | 13·5 | 12·5 | 11·4 |
| Standard deviation | 4·2 | 4·6 | 4·4 | 4·7 | 3·1 | 4·1 | 2·5 |
| Standard error | 0·5 | 0·6 | 0·6 | 0·6 | 0·4 | 0·7 | 0·5 |
| Range | 8–23 | 9–28 | 6–28 | 7–34 | 7–23 | 7–28 | 8–17 |
| Number of samples | 65 | 69 | 57 | 62 | 56 | 38 | 26 |

Males, p=not significant; females, p = <0·01.

TABLE 15

Number of school children of different ethnic origins by sex and area

| Area | Male A | B | C | D | Female A | B | C | D | Total |
|---|---|---|---|---|---|---|---|---|---|
| *Ethnic origin* | | | | | | | | | |
| *Caucasian* | | | | | | | | | |
| Number | 57 | 59 | 91 | 85 | 69 | 63 | 89 | 77 | 590 |
| Percentage (of total) | 8 | 8 | 12 | 12 | 9 | 9 | 12 | 11 | 81 |
| *West Indian* | | | | | | | | | |
| Number | 17 | 18 | 7 | 0 | 15 | 30 | 9 | 0 | 96 |
| Percentage | 2 | 2 | 1 | 0 | 2 | 4 | 1 | 0 | 13 |
| *Asian* | | | | | | | | | |
| Number | 5 | 17 | 1 | 1 | 3 | 17 | 1 | 0 | 45 |
| Percentage | 1 | 2 | | | | 2 | | 0 | 6 |
| *Unknown* | | | | | | | | | |
| Number | 1 | 0 | 0 | 0 | 0 | 0 | 0 | 0 | 1 |
| Percentage | | 0 | 0 | 0 | 0 | 0 | 0 | 0 | |
| *Total* | | | | | | | | | |
| Number | 80 | 94 | 99 | 86 | 87 | 110 | 99 | 77 | 732 |
| Percentage | 11 | 13 | 14 | 12 | 12 | 15 | 14 | 11 | 100 |

TABLE 16

Blood lead concentrations ($\mu$g/100 ml) of school children by ethnic origin

| Area | Male A | B | C | D | Female A | B | C | D | Total |
|---|---|---|---|---|---|---|---|---|---|
| *Ethnic origin* | | | | | | | | | |
| *Caucasian* | | | | | | | | | |
| Number of samples | 57 | 59 | 91 | 85 | 69 | 63 | 89 | 77 | 590 |
| Mean | 15·7 | 16·6 | 14·4 | 12·9 | 15·4 | 13·5 | 13·2 | 13·1 | 14·2 |
| Standard deviation | 4·6 | 5·0 | 3·6 | 3·1 | 4·3 | 3·5 | 3·7 | 3·7 | 4·1 |
| Standard error | 0·6 | 0·7 | 0·4 | 0·3 | 0·5 | 0·4 | 0·4 | 0·4 | 0·2 |
| Range | 7–30 | 9–32 | 7–23 | 7–22 | 8–29 | 8–26 | 6–26 | 7–28 | 6–32 |
| *West Indian* | | | | | | | | | |
| Number of samples | 17 | 18 | 7 | | 15 | 30 | 9 | | 96 |
| Mean | 18·1 | 14·8 | 16·4 | | 17·1 | 16·7 | 16·6 | | 16·6 |
| Standard deviation | 5·9 | 4·2 | 5·9 | | 6·5 | 5·0 | 4·5 | | 5·2 |
| Standard error | 1·4 | 1·0 | 2·2 | | 1·7 | 0·9 | 1·5 | | 0·5 |
| Range | 7–30 | 10–26 | 9–24 | | 9–34 | 9–28 | 10–22 | | 7–34 |
| *Asian* | | | | | | | | | |
| Number of samples | 5 | 17 | 1 | 1 | 3 | 17 | 1 | | 45 |
| Mean | 17·8 | 16·8 | 13·0 | 17·0 | 14·0 | 15·9 | 13·0 | | 16·2 |
| Standard deviation | 3·4 | 5·0 | | | 8·9 | 4·8 | | | 4·8 |
| Standard error | 1·5 | 1·2 | | | 5·1 | 1·2 | | | 0·7 |
| Range | 13–22 | 8–29 | | | 7–24 | 9–26 | | | 7–29 |
| *Unknown* | | | | | | | | | |
| Number of samples | 1 | | | | | | | | 1 |
| Mean | 23·0 | | | | | | | | 23·0 |
| *Total* | | | | | | | | | |
| Number of samples | 80 | 94 | 99 | 86 | 87 | 110 | 99 | 77 | 732 |
| Mean | 16·4 | 16·3 | 14·5 | 12·9 | 15·6 | 14·8 | 13·5 | 13·1 | 14·6 |
| Standard deviation | 4·9 | 4·9 | 3·8 | 3·1 | 4·9 | 4·4 | 3·8 | 3·7 | 4·4 |
| Standard error | 0·6 | 0·5 | 0·4 | 0·3 | 0·5 | 0·4 | 0·4 | 0·4 | 0·2 |
| Range | 7–30 | 8–32 | 7–24 | 7–22 | 7–34 | 8–28 | 6–26 | 7–28 | 6–34 |

## TABLE 17

Blood lead concentration ($\mu$g/100 ml) in school children by age of house

| | < 25 | 25–49 | 50–74 | 75+ | Unknown |
|---|---|---|---|---|---|
| | | Age of House, years | | | |
| *Males* | | | | | |
| Mean | 13·7 | 15·6 | 16·7 | 17·1 | 16·9 |
| Standard deviation | 3·8 | 4·6 | 4·5 | 6·1 | 5·5 |
| Standard error | 0·3 | 0·4 | 0·6 | 1·9 | 2·0 |
| Range | 7–28 | 7–31 | 9–32 | 10–30 | 9–23 |
| Number of samples | 157 | 129 | 55 | 10 | 8 |
| *Females* | | | | | |
| Mean | 13·3 | 14·8 | 15·2 | 15·7 | 16·7 |
| Standard deviation | 4·2 | 4·3 | 4·6 | 3·5 | 3·5 |
| Standard error | 0·3 | 0·4 | 0·6 | 1·0 | 0·9 |
| Range | 7–34 | 7–28 | 6–26 | 9–13 | 7–22 |
| Number of samples | 167 | 117 | 59 | 13 | 17 |

**TABLE 18**

Blood lead concentration ($\mu$g/100 ml) by lead pipes and age of house for school children

| Age of house, years | | Male | | | | | Female | | | | | Total |
|---|---|---|---|---|---|---|---|---|---|---|---|---|
| | | <25 | 25–49 | 50–74 | 75+ | Unknown | <25 | 25–49 | 50–74 | 75+ | Unknown | |
| *Lead pipes* | | | | | | | | | | | | |
| *Yes* | Mean | 14·3 | 15·7 | 16·6 | 17·3 | | 14·5 | 14·8 | 15·3 | 15·9 | 17·0 | 15·5 |
| | Standard deviation | 6·5 | 4·4 | 4·6 | 6·4 | | 6·6 | 4·2 | 4·5 | 3·6 | 0·0 | 4·5 |
| | Standard error | 2·5 | 0·4 | 0·7 | 2·1 | | 3·3 | 0·4 | 0·6 | 1·0 | 0·0 | 0·3 |
| | Range | 10–28 | 7–31 | 9–32 | 10–30 | | 8–23 | 8–28 | 6–26 | 9–12 | | 6–32 |
| | Number of samples | 7 | 109 | 44 | 9 | | 4 | 90 | 51 | 12 | 1 | 327 |
| *No* | Mean | 13·7 | 15·2 | 19·8 | | | 13·3 | 14·2 | 14·5 | 13·0 | 19·0 | 13·7 |
| | Standard deviation | 3·6 | 5·9 | 3·5 | | | 4·1 | 4·3 | 5·5 | 0·0 | 1·4 | 4·1 |
| | Standard error | 0·3 | 1·6 | 1·8 | | | 0·3 | 1·0 | 2·7 | 0·0 | 1·0 | 0·2 |
| | Range | 7–25 | 10–29 | 16–24 | | | 7–34 | 7–26 | 9–22 | | 18–20 | 7–34 |
| | Number of samples | 149 | 14 | 4 | | | 160 | 21 | 4 | 1 | 2 | 355 |
| *Unknown* | Mean | 11·0 | 14·8 | 15·7 | 15·0 | 16·9 | 12·0 | 16·0 | 15·0 | | 16·4 | 15·6 |
| | Standard deviation | 0·0 | 4·4 | 3·8 | 0·0 | 5·5 | 4·6 | 6·7 | 6·1 | | 3·7 | 4·6 |
| | Standard error | 0·0 | 1·8 | 1·4 | 0·0 | 2·0 | 2·7 | 2·7 | 3·0 | | 1·0 | 0·7 |
| | Range | | 11–22 | 12–23 | | 9–23 | 7–16 | 10–28 | 9–23 | | 7–22 | 7–28 |
| | Number of samples | 1 | 6 | 7 | 1 | 8 | 3 | 6 | 4 | | 14 | 50 |
| *Total* | Mean | 13·7 | 15·6 | 16·7 | 17·1 | 16·9 | 13·3 | 14·8 | 15·2 | 15·7 | 16·7 | 14·6 |
| | Standard deviation | 3·8 | 4·6 | 4·5 | 6·1 | 5·5 | 4·2 | 4·3 | 4·6 | 3·5 | 3·5 | 4·4 |
| | Standard error | 0·3 | 0·4 | 0·6 | 1·9 | 2·0 | 0·3 | 0·4 | 0·6 | 1·0 | 0·9 | 0·2 |
| | Range | 7–28 | 7–31 | 9–32 | 10–30 | 9–23 | 7–34 | 7–28 | 6–26 | 9–13 | 7–22 | 6–34 |
| | Number of samples | 157 | 129 | 55 | 10 | 8 | 167 | 117 | 59 | 13 | 17 | 732 |

## TABLE 19

Percentage of school children in each area with blood leads $\geqslant 25\mu g/100$ ml

| Area | | A | B | C | D | Total |
|---|---|---|---|---|---|---|
| *Females* | n | 3 | 4 | 1 | 1 | 9 |
| | N | 84 | 106 | 98 | 76 | 364 |
| | Percentage | 3·6 | 3·8 | 1·0 | 1·3 | 2·5 |
| *Males* | n | 4 | 7 | 0 | 0 | 11 |
| | N | 76 | 87 | 99 | 86 | 348 |
| | Percentage | 5·3 | 8·0 | 0·0 | 0·0 | 3·2 |

n = number with blood leads $\geqslant 25\mu g/100$ ml.
N = number with blood leads $< 25\mu g/100$ ml.

## TABLE 20

Blood lead concentrations $\geqslant 25\mu g/100$ ml in school children

| Area | | A | B | C | D |
|---|---|---|---|---|---|
| *Females* | C | 28(9); 29(11); | 26(9); | 26(9); | 28(13); |
| | N | 34(11); | 26(9); 28(10); 32(10); | | |
| | A | | | | |
| *Males* | C | 30(9); 25(13); | 27(8); 28(10); 32(10); 28(13); 31(14); | | |
| | N | 30(9); 25(10); | 26(10); | | |
| | A | | 29(10); | | |

Ages shown in brackets.
C = Caucasian   N = Negro   A = Asian

## TABLE 21

Blood lead concentrations ($\mu g/100$ ml) in children of lead workers

| Sex | Area | Age, years | Blood lead concentration | Father's occupation |
|---|---|---|---|---|
| M | A | 10 | 15 | Plumber |
| M | A | 10 | 18 | Painter |
| M | A | 11 | 13 | Car assembler |
| M | A | 12 | 20 | Lead battery assembler |
| M | A | 12 | 17 | Plumber |
| M | A | 12 | 22 | Machinist in battery factory |
| M | B | 9 | 24 | Machinist in battery factory |
| M | B | 11 | 15 | Pipe fitter in battery factory |
| M | B | 14 | 14 | Pipe fitter |
| M | C | 12 | 9 | Battery assembler |
| M | C | 12 | 21 | Printer |
| M | D | 11 | 22 | Printing Compositor |
| M | D | 11 | 15 | Plumber |
| F | A | 9 | 15 | Plumber |
| F | B | 9 | 24 | Battery assembler |
| F | C | 8 | 12 | Plumber |
| F | C | 9 | 14 | Enameler |
| F | C | 10 | 15 | Painter/decorator |
| F | C | 10 | 19 | Enameller |
| F | C | 11 | 13 | Sheet metal worker |
| F | C | 13 | 14 | Painter/decorator |
| F | D | 8 | 11 | Painter/decorator |

## TABLE 22

Blood lead concentration ($\mu$g/100 ml) in adults

|                     | Males | Females |
| ------------------- | ----- | ------- |
| Mean                | 22·1  | 14·7    |
| Standard deviation  | 6·4   | 5·0     |
| Standard error      | 0·4   | 0·3     |
| Range               | 9–46  | 3–35    |
| Number of samples   | 225   | 221     |

## TABLE 23

Occupational and environmental details of adults with blood lead $\geqslant 35\mu$g/100 ml, not working with lead

|         | Blood lead, $\mu$g/100 ml | Area (home address) | Lead pipes | Age of house, years | Occupation         |
| ------- | ------------------------- | ------------------- | ---------- | ------------------- | ------------------ |
| *Males* |                           |                     |            |                     |                    |
|         | 35                        | C                   | No         | 25–49               | Pipe fitter*       |
|         | 35                        | C                   | No         | < 25                | Office manager     |
|         | 35                        | A                   | No         | < 25                | Teacher            |
|         | 37                        | C                   | Yes        | 25–49               | Millwright         |
|         | 40                        | C                   | Yes        | 25–49               | Blacksmith         |
|         | 41                        | B                   | Yes        | 25–49               | Carpenter          |
|         | 42                        | B                   | Yes        | 75 +                | Assembler          |
|         | 43                        | C                   | No         | < 25                | Plant fitter       |
|         | 46                        | B                   | No         | 50–74               | Telephone engineer |
| *Females* |                         |                     |            |                     |                    |
|         | 35                        | C                   | Unknown    | 25–49               | Nurse              |

*note that this person, who was not occupationally exposed to lead, is not the same person as the pipe fitter in Table 29 with an identical blood lead concentration; the latter did claim an occupational exposure to lead.

## TABLE 24

Variation of blood lead concentration ($\mu$g/100 ml) with age of house, lead pipes and sex for adults

| Age of house, years | | Male | | | | | Female | | | | | | Total |
|---|---|---|---|---|---|---|---|---|---|---|---|---|---|
| | | <25 | 25-49 | 50-74 | 75+ | Unknown | <25 | 25-49 | 50-74 | 75+ | Unknown | | |
| *Lead pipes* | | | | | | | | | | | | | |
| Yes | Mean | 19·8 | 22·8 | 23·4 | 25·2 | 31·0 | 16·5 | 15·7 | 18·1 | 14·8 | 15·0 | | 20·0 |
| | Standard deviation | 6·6 | 6·6 | 7·0 | 10·5 | 0·0 | 6·4 | 3·3 | 5·0 | 3·7 | 0·0 | | 6·7 |
| | Standard error | 3·3 | 1·2 | 2·0 | 4·7 | 0·0 | 4·5 | 0·7 | 1·3 | 1·7 | 0·0 | | 0·7 |
| | Range | 10-24 | 13-41 | 12-34 | 17-42 | | 12-21 | 10-21 | 12-27 | 11-20 | | | 10-42 |
| | Number of samples | 4 | 32 | 12 | 5 | 1 | 2 | 24 | 14 | 5 | 1 | | 100 |
| No | Mean | 21·3 | 22·7 | 26·1 | 25·2 | 22·0 | 12·5 | 19·1 | 16·8 | 16·5 | 14·0 | | 19·8 |
| | Standard deviation | 6·5 | 5·7 | 9·2 | 6·5 | 0·0 | 4·3 | 6·2 | 4·0 | 3·5 | 0·0 | | 7·2 |
| | Standard error | 0·8 | 1·0 | 2·8 | 2·9 | 0·0 | 0·7 | 1·7 | 2·0 | 2·5 | 0·0 | | 0·5 |
| | Range | 9-43 | 12-35 | 12-46 | 18-33 | | 9-14 | 9-33 | 11-20 | 14-19 | | | 9-46 |
| | Number of samples | 74 | 33 | 11 | 5 | 1 | 37 | 13 | 4 | 2 | 1 | | 181 |
| *Unknown* | Mean | 19·9 | 21·6 | 20·2 | 16·0 | 21·0 | 12·4 | 14·9 | 14·6 | 18·5 | 13·9 | | 16·0 |
| | Standard deviation | 4·4 | 6·8 | 4·1 | 0·0 | 5·5 | 4·7 | 5·5 | 4·1 | 4·3 | 4·3 | | 5·8 |
| | Standard error | 1·1 | 2·3 | 1·4 | 0·0 | 1·7 | 0·9 | 0·8 | 1·1 | 1·8 | 1·0 | | 0·5 |
| | Range | 13-32 | 13-34 | 14-27 | | 15-33 | 3-22 | 8-35 | 10-24 | 13-25 | 7-24 | | 3-35 |
| | Number of samples | 17 | 9 | 9 | 1 | 11 | 31 | 47 | 14 | 6 | 20 | | 165 |
| *Total* | Mean | 21·0 | 22·6 | 23·4 | 24·4 | 21·9 | 12·6 | 15·8 | 16·4 | 16·8 | 14·0 | | 18·5 |
| | Standard deviation | 6·1 | 6·2 | 7·4 | 8·3 | 5·7 | 4·5 | 5·2 | 8·9 | 4·0 | 7·7 | | 6·8 |
| | Standard error | 0·6 | 0·7 | 0·6 | 2·5 | 1·6 | 0·5 | 0·6 | 1·2 | 1·1 | 1·2 | | 0·3 |
| | Range | 9-43 | 12-41 | 12-46 | 16-42 | 15-33 | 3-22 | 8-35 | 10-27 | 11-25 | 7-24 | | 3-46 |
| | Number of samples | 95 | 74 | 32 | 11 | 13 | 70 | 84 | 32 | 13 | 22 | | 446 |

**TABLE 25**

Variation of blood lead concentration ($\mu$g/100 ml) with age

| | Age range, years | | | | | | | | |
| | 15–19 | 20–24 | 25–29 | 30–34 | 35–39 | 40–44 | 45–49 | 50–64 | Unknown |
|---|---|---|---|---|---|---|---|---|---|
| *Males* | | | | | | | | | |
| Mean | 19·8 | 21·2 | 21·6 | 23·8 | 23·8 | 22·3 | 22·4 | 21·3 | 14·0 |
| Standard deviation | 4·7 | 7·5 | 6·6 | 6·7 | 6·4 | 4·7 | 5·1 | 6·6 | |
| Standard error | 1·3 | 1·2 | 0·9 | 1·1 | 1·3 | 1·2 | 1·1 | 1·5 | |
| Range | 13–29 | 9–46 | 13–43 | 10–41 | 14–42 | 16–30 | 14–32 | 11–37 | |
| Number of samples | 12 | 40 | 53 | 39 | 23 | 15 | 22 | 20 | 1 |
| *Females* | | | | | | | | | |
| Mean | 13·7 | 13·5 | 13·8 | 14·8 | 13·7 | 17·2 | 19·0 | 18·5 | 13·0 |
| Standard deviation | 4·6 | 4·6 | 4·4 | 4·2 | 4·4 | 5·0 | 6·6 | 5·0 | 0 |
| Standard error | 0·9 | 0·6 | 0·7 | 0·9 | 1·1 | 1·6 | 1·8 | 1·0 | 0 |
| Range | 7–24 | 3–24 | 6–27 | 7–22 | 4–22 | 11–25 | 11–35 | 11–33 | 13 |
| Number of samples | 25 | 68 | 44 | 20 | 15 | 10 | 13 | 24 | 2 |

**TABLE 26**

Variation of blood lead concentration ($\mu$g/100 ml) with smoking

| Number of Cigarettes/day | | Male | Female | Total |
|---|---|---|---|---|
| < 10 | Mean | 23·2 | 14·7 | 19·1 |
| | Standard deviation | 7·2 | 5·7 | 7·8 |
| | Standard error | 1·9 | 1·5 | 1·4 |
| | Range | 16–41 | 4–24 | 4–41 |
| | Number of samples | 15 | 14 | 29 |
| 10–19 | Mean | 22·0 | 14·4 | 19·2 |
| | Standard deviation | 6·5 | 5·3 | 7·1 |
| | Standard error | 1·0 | 1·1 | 0·9 |
| | Range | 12–46 | 4–24 | 4–26 |
| | Number of samples | 39 | 23 | 62 |
| > 20 | Mean | 24·4 | 16·3 | 21·6 |
| | Standard deviation | 6·8 | 5·0 | 7·3 |
| | Standard error | 1·0 | 1·0 | 0·9 |
| | Range | 13–43 | 3–27 | 3–43 |
| | Number of samples | 46 | 24 | 70 |
| Non-smokers | Mean | 21·2 | 14·6 | 17·5 |
| | Standard deviation | 6·0 | 4·9 | 6·3 |
| | Standard error | 0·5 | 0·4 | 0·4 |
| | Range | 9–42 | 6–35 | 6–42 |
| | Number of samples | 125 | 160 | 285 |


**TABLE 27**

Distribution of adults sampled by area and ethnic origin

| | Male | | | | Female | | | | Total |
|---|---|---|---|---|---|---|---|---|---|
| | A | B | C | D | A | B | C | D | |
| *ETHNIC ORIGIN* | | | | | | | | | |
| *Caucasian* | | | | | | | | | |
| Number | 11 | 45 | 150 | 17 | 11 | 58 | 137 | 14 | 443 |
| Percentage | 2 | 10 | 34 | 4 | 2 | 13 | 31 | 3 | 99 |
| *West Indian* | | | | | | | | | |
| Number | 0 | 1 | 1 | 0 | 0 | 1 | 0 | 0 | 3 |
| Percentage | 0 | | | 0 | 0 | | 0 | 0 | 1 |
| *Asian* | | | | | | | | | |
| Number | 0 | 0 | 0 | 0 | 0 | 0 | 0 | 0 | 0 |
| Percentage | 0 | 0 | 0 | 0 | 0 | 0 | 0 | 0 | 0 |
| *Total* | | | | | | | | | |
| Number | 11 | 46 | 151 | 17 | 11 | 59 | 137 | 14 | 446 |
| Percentage | 2 | 10 | 34 | 4 | 2 | 13 | 31 | 3 | 100 |

## TABLE 28

### Blood lead concentrations (μg/100 ml) of adults by sex and area

| Sex | Male | | | | Female | | | | Total |
|---|---|---|---|---|---|---|---|---|---|
| Area | A | B | C | D | A | B | C | D | |
| Number of samples | 11 | 46 | 151 | 17 | 11 | 59 | 137 | 14 | 446 |
| Mean | 23·5 | 22·4 | 22·2 | 19·7 | 14·1 | 15·6 | 14·5 | 14·0 | 18·5 |
| Standard deviation | 5·2 | 7·7 | 6·3 | 4·5 | 5·5 | 4·5 | 5·3 | 3·1 | 6·8 |
| Standard error | 1·6 | 1·1 | 0·5 | 1·1 | 1·7 | 0·6 | 0·5 | 0·8 | 0·3 |
| Range | 18–35 | 13–46 | 9–43 | 10–27 | 7–25 | 8–27 | 3–35 | 7–18 | 3–46 |

## TABLE 29

### Blood lead concentrations in lead workers

| Blood lead μg/100 ml | Occupation |
|---|---|
| 35 | Pipe fitter* |
| 35 | Plumber |
| 41 | Painter/decorator |
| 42 | Battery worker |
| 45 | Plumber |
| 48 | Battery worker |
| 51 | Battery worker |
| 52 | Battery worker |
| 62 | Battery worker |

*This subject was a different person from that in Table 23; the latter did not claim an occupational exposure to lead.

## TABLE 30

### Response to letters sent to parents of pre-school children

| | Number | Percentage |
|---|---|---|
| No response | 931 | 58·2 |
| Letters returned* | 270 | 16·9 |
| Permission refused | 46 | 2·9 |
| Permission granted | 353 | 22·0 |
| Total | 1600 | 100·0 |

*Reasons for return included: change of address; address unknown; house empty; house demolished!

## TABLE 31

### Total attendance of pre-school children

| | Areas | | | | |
|---|---|---|---|---|---|
| | A | B | C | D | Totals |
| *Males* | | | | | |
| Random | 5 | 15 | 38 | 63 | 121 |
| Non-random | 1 | 20 | 43 | 39 | 103 |
| Total | 6 | 35 | 81 | 102 | 224 |
| *Females* | | | | | |
| Random | 4 | 17 | 33 | 68 | 122 |
| Non-random | 1 | 22 | 27 | 44 | 94 |
| Total | 5 | 39 | 60 | 112 | 216 |

## TABLE 32

Blood lead concentrations ($\mu$g/100 ml) in pre-school children, by area

| Area | Male | | | | Female | | | | |
|---|---|---|---|---|---|---|---|---|---|
| | A | B | C | D | A | B | C | D | Total |
| *Random* | | | | | | | | | |
| Mean | 15·8 | 26·5 | 21·6 | 17·4 | 15·3 | 19·6 | 20·2 | 16·8 | 18·8 |
| Standard deviation | 6·1 | 16·7 | 12·8 | 4·7 | 3·3 | 4·8 | 8·1 | 5·3 | 8·4 |
| Standard error | 2·7 | 4·6 | 2·2 | 0·6 | 1·7 | 1·2 | 1·4 | 0·7 | 0·6 |
| Range | 8–24 | 10–62 | 10–89 | 8–33 | 11–19 | 11–29 | 8–49 | 9–36 | 8–89 |
| Number of samples | 5 | 13 | 35 | 60 | 4 | 17 | 32 | 65 | 231 |
| *Non-random* | | | | | | | | | |
| Mean | 15·0 | 25·3 | 22·8 | 21·2 | 16·0 | 25·6 | 20·7 | 17·7 | 21·6 |
| Standard deviation | — | 14·0 | 11·0 | 11·4 | — | 7·3 | 6·3 | 5·6 | 9·7 |
| Standard error | — | 3·1 | 1·6 | 1·9 | — | 1·6 | 1·2 | 0·8 | 0·7 |
| Range | — | 11–70 | 9–80 | 11–76 | — | 14–47 | 12–34 | 7–38 | 7–80 |
| Number of samples | 1 | 20 | 45 | 37 | 1 | 22 | 27 | 45 | 198 |
| *Total* | | | | | | | | | |
| Mean | 15·7 | 25·7 | 22·3 | 18·8 | 15·4 | 23·0 | 20·4 | 17·2 | 20·1 |
| Standard deviation | 5·4 | 14·9 | 11·8 | 8·1 | 2·9 | 6·9 | 7·3 | 5·4 | 9·1 |
| Standard error | 2·2 | 2·6 | 1·3 | 0·8 | 1·3 | 1·1 | 1·0 | 0·5 | 0·4 |
| Range | 8–24 | 10–70 | 9–89 | 8–76 | 11–19 | 11–47 | 8–49 | 7–38 | 7–89 |
| Number of samples | 6 | 33 | 80 | 97 | 5 | 39 | 59 | 110 | 429 |

## TABLE 33

Number of pre-school children with blood lead concentrations $\geqslant 35\mu g/100$ ml
and $\geqslant 25\mu g/100$ ml

| | | Male A | B | C | D | Female A | B | C | D | Total |
|---|---|---|---|---|---|---|---|---|---|---|
| *Random* | | | | | | | | | | |
| $\geqslant 35$ | Number | 0 | 2 | 1 | 0 | 0 | 0 | 1 | 1 | 5 |
| | Percentage | 0 | 15 | 3 | 0 | 0 | 0 | 3 | 2 | 2 |
| $\geqslant 25$ | Number | 0 | 3 | 5 | 4 | 0 | 3 | 7 | 4 | 26 |
| | Percentage | 0 | 23 | 14 | 7 | 0 | 18 | 22 | 6 | 11 |
| $< 25$ | Number | 5 | 8 | 29 | 56 | 4 | 14 | 24 | 60 | 200 |
| | Percentage | 100 | 62 | 83 | 93 | 100 | 82 | 75 | 92 | 87 |
| *Total* | | 5 | 13 | 35 | 60 | 4 | 17 | 32 | 65 | *231* |
| *Non-random* | | | | | | | | | | |
| $\geqslant 35$ | Number | 0 | 4 | 2 | 2 | 0 | 1 | 0 | 1 | 10 |
| | Percentage | 0 | 20 | 4 | 5 | 0 | 5 | 0 | 2 | 5 |
| $\geqslant 25$ | Number | 0 | 3 | 15 | 6 | 0 | 11 | 8 | 3 | 46 |
| | Percentage | 0 | 15 | 33 | 16 | 0 | 50 | 30 | 7 | 23 |
| $< 25$ | Number | 1 | 13 | 28 | 29 | 1 | 10 | 19 | 41 | 142 |
| | Percentage | 100 | 65 | 62 | 78 | 100 | 45 | 70 | 91 | 72 |
| *Total* | | 1 | 20 | 45 | 37 | 1 | 22 | 27 | 45 | *198* |
| *TOTAL* | | | | | | | | | | |
| $\geqslant 35$ | Number | 0 | 6 | 3 | 2 | 0 | 1 | 1 | 2 | 15 |
| | Percentage | 0 | 18 | 4 | 2 | 0 | 3 | 2 | 2 | 3 |
| $\geqslant 25$ | Number | 0 | 6 | 20 | 10 | 0 | 14 | 15 | 7 | 72 |
| | Percentage | 0 | 18 | 25 | 10 | 0 | 36 | 25 | 6 | 17 |
| $< 25$ | Number | 6 | 21 | 57 | 85 | 5 | 24 | 43 | 101 | 342 |
| | Percentage | 100 | 64 | 71 | 88 | 100 | 62 | 73 | 92 | 80 |
| *Total* | | 6 | 33 | 80 | 97 | 5 | 39 | 59 | 110 | *429* |

## TABLE 34

Variation of blood lead concentration ($\mu g/100$ ml) with age

| Age, years | Male 1 | 2 | 3 | 4 | Female 1 | 2 | 3 | 4 |
|---|---|---|---|---|---|---|---|---|
| *Random* | | | | | | | | |
| Mean | 19·1 | 20·2 | 22·1 | 16·9 | 16·0 | 20·2 | 18·0 | 17·2 |
| Standard deviation | 8·9 | 12·9 | 10·1 | 5·2 | 4·4 | 7·7 | 5·2 | 6·0 |
| Standard error | 1·6 | 2·1 | 2·1 | 1·1 | 0·9 | 1·3 | 1·0 | 1·0 |
| Range | 8–57 | 10–89 | 14–62 | 10–32 | 9–25 | 11–49 | 10–30 | 8–29 |
| Number of samples | 31 | 37 | 23 | 22 | 22 | 34 | 28 | 34 |
| *Non-random* | | | | | | | | |
| Mean | 23·1 | 22·0 | 21·7 | 23·3 | 17·9 | 21·9 | 21·3 | 20·3 |
| Standard deviation | 14·2 | 14·4 | 12·0 | 8·1 | 5·6 | 8·0 | 5·3 | 7·2 |
| Standard error | 3·0 | 3·2 | 2·5 | 1·3 | 1·3 | 1·7 | 1·4 | 1·2 |
| Range | 11–76 | 13–80 | 9–70 | 12–42 | 7–26 | 9–34 | 14–34 | 12–47 |
| Number of samples | 22 | 21 | 23 | 37 | 19 | 23 | 15 | 38 |
| *Total* | | | | | | | | |
| Mean | 20·8 | 20·8 | 21·9 | 20·9 | 16·9 | 20·9 | 19·2 | 18·8 |
| Standard deviation | 11·4 | 13·4 | 11·0 | 7·8 | 5·0 | 7·8 | 5·4 | 6·8 |
| Standard error | 1·6 | 1·8 | 1·6 | 1·0 | 0·8 | 1·0 | 0·8 | 0·8 |
| Range | 8–76 | 10–89 | 9–70 | 10–42 | 7–26 | 9–49 | 10–34 | 8–47 |
| Number of samples | 53 | 58 | 46 | 59 | 41 | 57 | 43 | 72 |

## TABLE 35

Variation of blood lead concentrations ($\mu$g/100 ml) with age of house for pre-school children

| Age of house, years | Males <25 | 25–49 | 50–74 | 75+ | Unknown | Females <25 | 25–49 | 50–74 | 75+ | Unknown |
|---|---|---|---|---|---|---|---|---|---|---|
| *Random* | | | | | | | | | | |
| Mean | 18·8 | 19·3 | 19·9 | 29·3 | 32·5 | 15·9 | 20·4 | 19·0 | 19·5 | 21·3 |
| Standard deviation | 12·7 | 4·7 | 6·4 | 24·1 | 0·7 | 4·9 | 7·4 | 6·9 | 4·9 | 3·1 |
| Standard error | 1·8 | 0·7 | 1·6 | 13·9 | 0·5 | 0·6 | 1·2 | 2·2 | 2·0 | 1·3 |
| Range | 8–89 | 10–34 | 12–31 | 13–57 | 32–33 | 8–36 | 10–49 | 11–31 | 16–29 | 17–26 |
| Number of samples | 52 | 40 | 16 | 3 | 2 | 58 | 38 | 10 | 6 | 6 |
| *Non-random* | | | | | | | | | | |
| Mean | 20·2 | 22·2 | 20·5 | 37·6 | 22·6 | 17·5 | 20·2 | 21·5 | 29·3 | 25·8 |
| Standard deviation | 12·1 | 7·2 | 6·0 | 21·8 | 6·7 | 5·2 | 5·5 | 6·9 | 10·2 | 5·6 |
| Standard error | 2·0 | 1·3 | 1·6 | 7·3 | 2·0 | 4·3 | 1·0 | 1·9 | 3·6 | 2·3 |
| Range | 11–80 | 9–35 | 13–31 | 15–76 | 11–33 | 7–30 | 12–33 | 10–34 | 16–47 | 18–34 |
| Number of samples | 38 | 31 | 14 | 9 | 11 | 39 | 29 | 13 | 8 | 6 |
| *Total* | | | | | | | | | | |
| Mean | 19·4 | 20·6 | 20·2 | 35·5 | 24·2 | 16·5 | 20·3 | 20·4 | 25·1 | 23·6 |
| Standard deviation | 12·4 | 6·1 | 6·1 | 21·5 | 7·1 | 5·0 | 6·6 | 6·9 | 9·5 | 4·9 |
| Standard error | 1·3 | 0·7 | 1·1 | 6·2 | 2·0 | 0·1 | 0·8 | 1·4 | 2·5 | 1·4 |
| Range | 8–89 | 9–35 | 12–31 | 13–76 | 11–33 | 7–36 | 10–49 | 10–34 | 16–47 | 17–34 |
| Number of samples | 90 | 71 | 30 | 12 | 13 | 97 | 67 | 23 | 14 | 12 |

## TABLE 36

Variation of blood lead concentrations ($\mu$g/100 ml) with presence of lead pipes for pre-school children

| | Lead pipes in house | | |
| | Yes | No | Not known |
| --- | --- | --- | --- |
| *Random* | | | |
| Mean | 20·0 | 18·1 | 19·6 |
| Standard deviation | 4·7 | 9·3 | 7·9 |
| Standard error | 0·9 | 0·9 | 0·9 |
| Range | 13–32 | 8–89 | 8–49 |
| Number of samples | 28 | 120 | 83 |
| *Non-random* | | | |
| Mean | 21·8 | 20·5 | 22·8 |
| Standard deviation | 7·5 | 9·7 | 10·5 |
| Standard error | 1·4 | 1·0 | 1·2 |
| Range | 12–35 | 7–80 | 9–76 |
| Number of samples | 29 | 93 | 76 |
| *Combined* | | | |
| Mean | 20·9 | 19·1 | 21·1 |
| Standard deviation | 6·3 | 9·5 | 9·4 |
| Standard error | 0·8 | 0·7 | 0·7 |
| Range | 12–35 | 7–89 | 8–76 |
| Number of samples | 57 | 213 | 159 |


## TABLE 37a

Variation of blood lead concentration ($\mu$g/100 ml) with history of pica. Random sample of pre-school children

| | History of pica | |
| | Positive | Negative |
| --- | --- | --- |
| *Males* | | |
| Mean | 19·8 | 19·5 |
| Standard deviation | 12·3 | 7·6 |
| Standard error | 1·6 | 1·0 |
| Range | 8–89 | 9–62 |
| Number of samples | 56 | 57 |
| *Females* | | |
| Mean | 18·1 | 18·0 |
| Standard deviation | 6·3 | 6·2 |
| Standard error | 0·8 | 0·8 |
| Range | 9–49 | 8–36 |
| Number of samples | 58 | 60 |
| *Total* | | |
| Mean | 18·9 | 18·8 |
| Standard deviation | 10·1 | 6·9 |
| Standard error | 0·9 | 0·6 |
| Range | 8–89 | 8–62 |
| Number of samples | 114 | 117 |

## TABLE 37b

Variation of blood lead concentration ($\mu$g/100 ml) with history of pica.
Non-random sample of pre-school children

| | History of pica Positive | Negative |
|---|---|---|
| *Males* | | |
| Mean | 24·2 | 20·4 |
| Standard deviation | 14·0 | 6·6 |
| Standard error | 1·8 | 1·0 |
| Range | 11–80 | 9–42 |
| Number of samples | 61 | 42 |
| *Females* | | |
| Mean | 20·1 | 20·6 |
| Standard deviation | 6·3 | 7·4 |
| Standard error | 1·0 | 1·0 |
| Range | 7–38 | 10–47 |
| Number of samples | 43 | 52 |
| *Total* | | |
| Mean | 22·5 | 20·5 |
| Standard deviation | 11·6 | 7·0 |
| Standard error | 1·1 | 0·7 |
| Range | 7–80 | 9–47 |
| Number of samples | 104 | 94 |


## TABLE 37c

Variation of blood lead concentration ($\mu$g/100 ml) with history of pica.
Total sample of pre-school children

| | History of pica Positive | Negative |
|---|---|---|
| *Males* | | |
| Mean | 22·1 | 19·9 |
| Standard deviation | 13·3 | 7·2 |
| Standard error | 1·2 | 0·7 |
| Range | 8–89 | 9–62 |
| Number of samples | 117 | 99 |
| *Females* | | |
| Mean | 18·9 | 19·2 |
| Standard deviation | 6·4 | 6·9 |
| Standard error | 0·6 | 0·7 |
| Range | 7–49 | 8–47 |
| Number of samples | 101 | 112 |
| *Total* | | |
| Mean | 20·6 | 19·5 |
| Standard deviation | 10·8 | 7·0 |
| Standard error | 0·7 | 0·5 |
| Range | 7–89 | 8–62 |
| Number of samples | 218 | 211 |

**TABLE 38**

Blood leads of children whose father's occupation involved contact with lead

| Sex | Area | Blood lead, $\mu$g/100 ml | Occupation |
|-----|------|---------------------------|------------|
| M | B | 24 | Painter/decorator |
| M | C | 18 | ditto |
| M | C | 24 | ditto |
| F | D | 18 | ditto |
| F | C | 16 | ditto |
| F | D | 21 | ditto |
| F | B | 26 | ditto |

## TABLE 39

Follow-up studies

| Sex | Random | Age | Area | Lead pipes | Age of house, years | Ethnic‡ origin | Pica | Father's occupation | Blood lead, μg/100 ml 1 | 2 | 3 |
|---|---|---|---|---|---|---|---|---|---|---|---|
| M | ✓ | 3 | B | No | 3 | A | No | Baker | 62 | — | — |
| M | — | 4 | B | Yes | 20 | A | No | Labourer | 42 | 38 | 37 |
| M | — | 1 | B | Not known | 80 | C | Yes | Unemployed | 37 | 24 | 23 |
| *M | ✓ | 2 | B | Not known | 90¶ | A | Yes | Stamper | 57 | 50 | 54 |
| *F | — | 3 | B | Not known | 90¶ | A | Yes | Stamper | 70 | 51 | 50 |
| †M | — | 4 | B | No | 85 | A | Yes | Greengrocer | 38 | 38 | 33 |
| †F | — | 4 | B | No | 85 | A | No | Baker | 47 | 36 | 29 |
| †M | — | 6 | B | No | 85 | A | No | Baker | 40 | 29 | 33 |
| M | ✓ | 11 | B | Not known | Not known | C | Yes | Unemployed | 37 | 40 | 31 |
| F | — | 2 | C | Not known | 25–49 | C | Yes | Lorry driver | 49 | 43 | 33 |
| §M | ✓ | 2 | C | No | 20 | C | No | Service engineer | 80 | 23 | 17 |
| §M | ✓ | 2 | C | No | 20 | C | No | Service engineer | 89 | 18 | 15 |
| M | — | 1 | C | Yes | 40 | C | Yes | Electrician | 35 | 29 | 27 |
| F | ✓ | 2 | D | No | 6 | C | No | Bank manager | 36 | 16 | 15 |
| M | — | 4 | D | No | 15 | C | Yes | Operation manager | 38 | 37 | 31 |
| **M | — | 1 | D | No | 100 | C | Yes | Engineer | 76 | 58 | 52 |
| **F | — | 4 | D | No | 100 | C | Yes | Engineer | 38 | 78 | 35 |

*Brother and sister.
†These three children lived in the same house; the second and third were brother and sister.
‡A = Asian; C = Caucasian.
§Identical twins.
¶House in a very poor state of repair with poor paintwork.
**Brother and sister.
1: First result from City Scientific Officer; 2: Second result from City Scientific Officer; 3: Result from Institute of Child Health.

## TABLE 40

Blood lead concentrations in residents living near the M6–A38(M) interchange
Capillary blood first series (μg/100 ml)

| | Area 1 | 2 | 3 | Total |
|---|---|---|---|---|
| *Males* | | | | |
| Mean | 13·8 | 15·8 | 17·6 | 16·0 |
| Standard deviation | 8·8 | 12·6 | 11·9 | 11·8 |
| Standard error | 1·1 | 1·0 | 1·1 | 0·6 |
| Number of samples | 68 | 173 | 118 | 359 |
| Range | 6–48 | 6–110 | 3–64 | 3–110 |
| *Females* | | | | |
| Mean | 10·5 | 12·2 | 14·7 | 12·7 |
| Standard deviation | 6·6 | 8·3 | 9·6 | 8·5 |
| Standard error | 0·6 | 0·5 | 0·8 | 0·4 |
| Number of samples | 78 | 225 | 139 | 442 |
| Range | 6–29 | 4–40 | 4–40 | 4–40 |
| *Children under 10* | | | | |
| Mean | 12·4 | 19·0 | 15·5 | 16·1 |
| Standard deviation | 6·0 | 13·3 | 12·0 | 11·6 |
| Standard error | 1·2 | 2·1 | 2·1 | 1·2 |
| Number of samples | 25 | 40 | 33 | 98 |
| Range | 6–25 | 6–58 | 6–58 | 6–58 |

## TABLE 41

Blood lead concentrations in residents living near the M6–A38(M) interchange
Venous blood second series (μg/100 ml)

| | Area 1 | 2 | 3 | Total |
|---|---|---|---|---|
| *Males* | | | | |
| Mean | 20·1 | 19·3 | 17·4 | 18·8 |
| Standard deviation | 8·5 | 7·0 | 6·4 | 7·2 |
| Standard error | 1·0 | 0·5 | 0·6 | 0·4 |
| Number of samples | 68 | 175 | 118 | 361 |
| Range | 6–46 | 7–42 | 7–42 | 6–46 |
| *Females* | | | | |
| Mean | 16·9 | 16·6 | 16·6 | 16·6 |
| Standard deviation | 6·5 | 6·1 | 5·9 | 6·1 |
| Standard error | 0·7 | 0·4 | 0·5 | 0·3 |
| Number of samples | 78 | 226 | 139 | 443 |
| Range | 5–46 | 4–32 | 4–36 | 4–36 |
| *Children under 10* | | | | |
| Mean | 17·7 | 17·8 | 15·3 | 16·9 |
| Standard deviation | 6·3 | 6·3 | 6·2 | 6·3 |
| Standard error | 1·2 | 1·0 | 1·1 | 0·6 |
| Number of samples | 25 | 40 | 33 | 98 |
| Range | 8–33 | 6–32 | 6–33 | 6–33 |

## TABLE 42

Comparison between paired capillary samples* ($\mu$g/100 ml)

| | Area 1 Series 1 | Series 2 | 2 Series 1 | Series 2 | 3 Series 1 | Series 2 |
|---|---|---|---|---|---|---|
| **Males** | | | | | | |
| Mean | 17·8 | 18·0 | 20·7 | 27·7 | 19·1 | 22·4 |
| Standard deviation | 9·5 | 8·1 | 10·9 | 10·9 | 10·1 | 10·5 |
| Standard error | 1·9 | 1·6 | 2·4 | 2·4 | 1·5 | 1·5 |
| Number of samples | 26 | 26 | 21 | 21 | 47 | 47 |
| Range | 7–39 | 7–33 | 8–48 | 8–50 | 7–48 | 7–50 |
| Significance | | ns† | | $P<0.05$ | | ns |
| **Females** | | | | | | |
| Mean | 12·2 | 13·2 | 15·8 | 22·2 | 13·6 | 16·7 |
| Standard deviation | 6·7 | 4·8 | 10·3 | 8·3 | 8·3 | 7·7 |
| Standard error | 1·2 | 0·9 | 2·4 | 1·9 | 1·2 | 1·1 |
| Number of samples | 30 | 30 | 19 | 19 | 49 | 49 |
| Range | 6–25 | 8–30 | 4–40 | 14–45 | 14–40 | 8–45 |
| Significance | | ns | | $P<0.01$ | | ns |
| **Children** | | | | | | |
| Mean | 12·3 | 21·0 | 15 | 20 | 13·0 | 20·7 |
| Standard deviation | 4·9 | 8·5 | | | 4·2 | 7·0 |
| Standard error | 2·8 | 4·9 | | | 2·1 | 3·5 |
| Number of samples | 3 | 3 | 1 | 1 | 4 | 4 |
| Range | 9–18 | 13–30 | | | 9–18 | 13–30 |
| Significance | | ns | | | | ns |

*Analysed at Regional Toxicology Laboratory.
†ns indicates "not significant".


## TABLE 43

Comparison of differences in Series 1 and 2 in paired capillary samples
and capillary – venous samples ($\mu$g/100 ml)

| | Area 1 | 2 | 3 |
|---|---|---|---|
| **Males** | | | |
| (cap-ven)[a] | +6·3 | +3·5 | −0·2 |
| (cap-cap)[b] | +0·2 | +7·0 | +3·3 |
| **Females** | | | |
| (cap-ven)[a] | +6·4 | +4·4 | +1·9 |
| (cap-cap)[b] | +1·0 | +6·4 | +3·1 |

[a]Means of total sample.
[b]Means of paired capillary samples.

## TABLE 44

Blood lead concentrations in residents living near the M6–A38(M) interchange
Venous blood third series* ($\mu$g/100 ml)

| | Area 1 | 2 | 3 | Total |
|---|---|---|---|---|
| *Males* | | | | |
| Mean | 29·4 | 30·3 | 29·2 | 29·8 |
| Standard deviation | 8·1 | 6·9 | 6·2 | 6·9 |
| Standard error | 1·1 | 0·6 | 0·6 | 0·4 |
| Number of samples | 49 | 128 | 93 | 270 |
| Range | 17–54 | 17–56 | 16–56 | 16–50 |
| *Females* | | | | |
| Mean | 25·7 | 24·7 | 25·6 | 25·2 |
| Standard deviation | 6·8 | 5·8 | 6·3 | 6·1 |
| Standard error | 0·9 | 0·4 | 0·6 | 0·3 |
| Number of samples | 59 | 176 | 108 | 343 |
| Range | 14–54 | 14–42 | 15–54 | 14–54 |
| *Children under 10* | | | | |
| Mean | 24·2 | 27·9 | 26·4 | 26·3 |
| Standard deviation | 3·2 | 7·8 | 5·3 | 6·1 |
| Standard error | 0·7 | 1·5 | 1·2 | 0·7 |
| Number of samples | 20 | 27 | 19 | 66 |
| Range | 17–29 | 17–60 | 19–41 | 17–60 |

*Analysed by City Analyst.

## TABLE 45

Summary of apparent* changes in mean blood lead concentrations in the
three series of samples ($\mu$g/100 ml)

| | | Area 1 | 2 | 3 | Total |
|---|---|---|---|---|---|
| *Males* | | | | | |
| Series change | 1:2 | + 6·3 | + 3·5 | − 0·2 | + 2·8 |
| | 2:3 | + 9·3 | +11·0 | +11·8 | +11·0 |
| | 1:3 | +15·6 | +14·5 | +11·6 | +13·8 |
| *Females* | | | | | |
| Series change | 1:2 | + 6·4 | + 4·4 | + 1·9 | + 3·9 |
| | 2:3 | + 8·8 | + 8·1 | + 9·0 | + 8·6 |
| | 1:3 | +15·2 | +12·5 | +10·9 | +12·5 |
| *Children* | | | | | |
| Series change | 1:2 | + 5·3 | − 1·2 | − 0·2 | + 0·8 |
| | 2:3 | + 6·5 | +10·1 | +11·1 | + 9·4 |
| | 1:3 | +11·8 | + 8·9 | +10·9 | +10·2 |

*Changes in this table take no account of the confounding factors imposed by change of sampling
technique and analytical laboratory and reference should be made to the text of Appendix C for
an interpretation.

## TABLE 46a

Mean blood lead concentrations in male subjects in all three series ($\mu$g/100 ml)

| | Area 1 | 2 | 3 | Total |
|---|---|---|---|---|
| *Series 1 – capillary*[a] | | | | |
| Mean | 14·4 | 15·9 | 16·9 | 16·2 |
| Standard deviation | 8·0 | 15·7 | 11·8 | 12·4 |
| Standard error | 1·1 | 1·2 | 1·2 | 0·7 |
| Number of samples | 49 | 127 | 93 | 269 |
| Range | 6–29 | 6–110 | 3–64 | 3–110 |
| *Series 2 – venous*[a] | | | | |
| Mean | 19·8 | 19·0 | 17·4 | 18·6 |
| Standard deviation | 8·7 | 6·9 | 6·5 | 7·2 |
| Standard error | 1·2 | 0·6 | 0·7 | 0·4 |
| Number of samples | 49 | 127 | 93 | 269 |
| Range | 6–46 | 5–40 | 7–42 | 5–42 |
| *Series 3 – venous*[b] | | | | |
| Mean | 29·4 | 30·3 | 29·2 | 29·8 |
| Standard deviation | 8·1 | 6·9 | 6·2 | 6·9 |
| Standard error | 1·1 | 0·6 | 0·6 | 0·4 |
| Number of samples | 49 | 127 | 93 | 269 |
| Range | 17–54 | 17–56 | 16–50 | 16–56 |

[a]Analysed at Regional Toxicology Laboratory.
[b]Analysed by City Analyst.

## TABLE 46b

Mean blood lead concentrations in female subjects in all three series ($\mu$g/100 ml)

| | Area 1 | 2 | 3 | Total |
|---|---|---|---|---|
| *Series 1 – capillary*[a] | | | | |
| Mean | 11·0 | 12·0 | 14·7 | 12·7 |
| Standard deviation | 6·0 | 8·2 | 9·8 | 8·5 |
| Standard error | 0·8 | 0·6 | 0·9 | 0·5 |
| Number of samples | 59 | 176 | 108 | 343 |
| Range | 6–28 | 4–39 | 4–37 | 4–39 |
| *Series 2 – venous*[a] | | | | |
| Mean | 16·4 | 16·4 | 16·9 | 16·4 |
| Standard deviation | 6·2 | 5·8 | 5·8 | 5·8 |
| Standard error | 0·8 | 0·4 | 0·6 | 0·3 |
| Number of samples | 59 | 176 | 108 | 343 |
| Range | 5–46 | 4–32 | 6–36 | 4–46 |
| *Series 3 – venous*[b] | | | | |
| Mean | 25·7 | 24·7 | 25·6 | 25·2 |
| Standard deviation | 6·8 | 5·8 | 6·3 | 6·1 |
| Standard error | 0·9 | 0·4 | 0·6 | 0·3 |
| Number of samples | 59 | 176 | 108 | 343 |
| Range | 14–54 | 14–42 | 15–54 | 14–54 |

[a]Analysed at Regional Toxicology Laboratory.
[b]Analysed by City Analyst.

### TABLE 46c

Mean blood lead concentrations in children in all three series ($\mu$g/100 ml)

| | Area 1 | 2 | 3 | Total |
|---|---|---|---|---|
| _Series 1 – capillary_[a] | | | | |
| Mean | 11·1 | 20·2 | 15·6 | 16·1 |
| Standard deviation | 5·9 | 13·8 | 10·4 | 11·4 |
| Standard error | 1·3 | 2·6 | 2·4 | 1·4 |
| Number of samples | 20 | 27 | 19 | 66 |
| Range | 8–24 | 8–25 | 6–25 | 6–25 |
| _Series 2 – venous_[a] | | | | |
| Mean | 16·7 | 16·3 | 15·4 | 16·2 |
| Standard deviation | 5·4 | 5·8 | 5·9 | 5·6 |
| Standard error | 1·2 | 1·1 | 1·3 | 0·7 |
| Number of samples | 20 | 27 | 19 | 66 |
| Range | 8–24 | 8–25 | 6–25 | 6–25 |
| _Series 3 – venous_[b] | | | | |
| Mean | 24·2 | 27·9 | 26·4 | 26·3 |
| Standard deviation | 3·2 | 7·8 | 5·3 | 6·1 |
| Standard error | 0·7 | 1·5 | 1·2 | 0·7 |
| Number of samples | 20 | 27 | 19 | 66 |
| Range | 17–29 | 17–60 | 19–41 | 17–60 |

[a]Analysed at Regional Toxicology Laboratory.
[b]Analysed by City Analyst.


### TABLE 47

Summary of apparent* differences in mean blood lead concentrations from subjects in all three series ($\mu$g/100 ml)

| | | Area 1 | 2 | 3 | Total |
|---|---|---|---|---|---|
| _Males_ | | | | | |
| Series change | 1:2 | + 5·4 | + 3·1 | + 0·5 | + 2·4 |
| | 2:3 | + 9·6 | +11·3 | +11·8 | +11·2 |
| | 1:3 | +15·0 | +14·4 | +12·3 | +13·6 |
| _Females_ | | | | | |
| Series change | 1:2 | + 5·4 | + 4·4 | + 2·2 | + 3·7 |
| | 2:3 | + 9·3 | + 8·3 | + 8·7 | + 8·8 |
| | 1:3 | +14·7 | +12·7 | +10·9 | +12·5 |
| _Children_ | | | | | |
| Series change | 1:2 | + 5·6 | − 3·9 | − 0·2 | + 0·1 |
| | 2:3 | + 7·5 | +11·6 | +11·0 | +10·1 |
| | 1:3 | +13·1 | + 7·7 | +10·8 | +10·2 |

*Changes in this table take no account of confounding factors and reference should be made to the text of Appendix C for an interpretation.

## TABLE 48

Blood lead in subjects in all three series, all analysed at Regional Toxicology Laboratory ($\mu$g/100 ml)

| | Series 1c | 2v | 3v |
|---|---|---|---|
| *Males* | | | |
| Mean | 14·4 | 18·4 | 23·7 |
| Standard deviation | 9·0 | 6·6 | 6·8 |
| Standard error | 1·4 | 1·0 | 1·1 |
| Number of samples | 41 | 41 | 41 |
| Range | 6–39 | 5–33 | 13–36 |
| | P 1:2<0·05   2:3<0·001 | 1:3<0·001 | |
| *Females* | | | |
| Mean | 10·8 | 15·1 | 19·1 |
| Standard deviation | 6·9 | 5·4 | 4·9 |
| Standard error | 0·9 | 0·7 | 0·6 |
| Number of samples | 58 | 58 | 58 |
| Range | 4–36 | 4–30 | 12–31 |
| | P 1:2<0·001   2:3<0·001 | 1:3<0·001 | |

c=capillary blood; v=venous blood.


## TABLE 49

Blood lead concentrations in October 1973 and October 1974 ($\mu$g/100 ml)

| | Males October 1973 | October 1974 | Females October 1973 | October 1974 |
|---|---|---|---|---|
| Mean | 29·6 | 28·8 | 24·2 | 23·2 |
| Standard deviation | 5·6 | 7·7 | 5·4 | 6·2 |
| Standard error | 0·9 | 1·3 | 0·8 | 0·9 |
| Number of samples | 38 | 38 | 53 | 53 |
| Range | 19–44 | 18–54 | 14–34 | 13–39 |

Note: Blood analysed by the City Analyst on *both* occasions. For this reason the October 1973 means are higher than shown for series 3 in Table 48.

# TABLE 50

Results of duplicate analyses on venous blood from school children
(lead concentrations in $\mu g/100$ ml)

| Scientific Officer | Institute of Child Health | Scientific Officer | Institute of Child Health | Scientific Officer | Institute of Child Health |
|---|---|---|---|---|---|
| 30 | 35 | 10 | 12 | 11 | 17 |
| 16 | 19 | 14 | 17 | 15 | 19 |
| 11 | 15 | 13 | 11 | 14 | 21 |
| 8 | 11 | 13 | 16 | 16 | 15 |
| 15 | 18 | 10 | 13 | 17 | 21 |
| 15 | 18 | 16 | 17 | 14 | 23 |
| 22 | 27 | 16 | 14 | 12 | 18 |
| 12 | 15 | 15 | 17 | 10 | 13 |
| 16 | 20 | 16 | 18 | 14 | 18 |
| 18 | 18 | 18 | 16 | 13 | 16 |
| 13 | 16 | 21 | 20 | 21 | 20 |
| 17 | 21 | 12 | 12 | 14 | 12 |
| 23 | 25 | 9 | 15 | 18 | 19 |
| 13 | 20 | 28 | 18 | 14 | 16 |
| 24 | 26 | 16 | 22 | 12 | 20 |
| 16 | 15 | 14 | 12 | 14 | 15 |
| 14 | 13 | 10 | 20 | 16 | 18 |
| 14 | 14 | 17 | 24 | 13 | 16 |
| 9 | 15 | 16 | 16 | 14 | 15 |
| 11 | 14 | 13 | 16 | 13 | 16 |
| 19 | 16 | 24 | 23 | 13 | 12 |
| 16 | 16 | 12 | 15 | 13 | 8 |
| 25 | 17 | 16 | 13 | 8 | 10 |
| 17 | 17 | 14 | 14 | 13 | 14 |
| 19 | 19 | 19 | 19 | 14 | 15 |
| 27 | 24 | 15 | 17 | 10 | 12 |
| 12 | 13 | 11 | 18 | 11 | 12 |
| 17 | 16 | 23 | 27 | 13 | 14 |
| 24 | 26 | 11 | 14 | 17 | 16 |
| 15 | 19 | 13 | 15 | 15 | 13 |
| 11 | 19 | 13 | 16 | 13 | 14 |
| 15 | 19 | 12 | 12 | 13 | 14 |
| 15 | 17 | 12 | 14 | 14 | 12 |
| 28 | 25 | 12 | 18 | 14 | 13 |
| 11 | 18 | 18 | 19 | 14 | 15 |
| 32 | 27 | 15 | 14 | 17 | 17 |
| 16 | 18 | 15 | 15 | 12 | 13 |
| 18 | 19 | 13 | 16 | 18 | 20 |
| 11 | 13 | 9 | 18 | 10 | 11 |
| 19 | 17 | 11 | 17 | 11 | 16 |
| 15 | 16 | 17 | 16 | 13 | 17 |
| 11 | 15 | 19 | 21 | 22 | 19 |
| 10 | 18 | 26 | 28 | 13 | 16 |
| 14 | 17 | 15 | 20 | 22 | 29 |
| 11 | 15 | 13 | 14 | 16 | 21 |
| 8 | 15 | 14 | 20 | 12 | 13 |
| 8 | 13 | 13 | 16 | 15 | 18 |
| 15 | 15 | 20 | 24 | 19 | 18 |
| 12 | 14 | 28 | 30 | 14 | 16 |
| 23 | 22 | 9 | 9 | 12 | 19 |
| 11 | 17 | 14 | 21 | 10 | 14 |
| 9 | 13 | 11 | 13 | 10 | 14 |
| 11 | 14 | 20 | 27 | 15 | 21 |
| 14 | 15 | 17 | 16 | 18 | 19 |
| 18 | 16 | 9 | 14 | 16 | 23 |
| 15 | 16 | 18 | 23 | 18 | 20 |
| 17 | 20 | 18 | 18 | 16 | 18 |
| 24 | 21 | 12 | 18 | 17 | 17 |
| 11 | 14 | 13 | 14 | 11 | 17 |
| 11 | 12 | 14 | 15 | 13 | 20 |
| 13 | 11 | 17 | 18 | 12 | 15 |

TABLE 50—*continued*

| Scientific Officer | Institute of Child Health | Scientific Officer | Institute of Child Health | Scientific Officer | Institute of Child Health |
|---|---|---|---|---|---|
| 18 | 17 | 15 | 15 | 15 | 16 |
| 15 | 15 | 14 | 19 | 11 | 12 |
| 12 | 14 | 11 | 14 | 10 | 14 |
| 15 | 14 | 16 | 14 | 8 | 7 |
| 22 | 21 | 11 | 12 | 9 | 10 |
| 19 | 21 | 13 | 11 | 14 | 16 |
| 13 | 13 | 16 | 11 | 13 | 16 |
| 34 | 27 | 19 | 16 | 9 | 7 |
| 22 | 19 | 9 | 10 | 10 | 11 |
| 14 | 18 | 16 | 15 | 18 | 20 |
| 8 | 13 | 13 | 11 | 10 | 12 |
| 13 | 20 | 13 | 14 | 13 | 15 |
| 10 | 13 | 10 | 11 | 11 | 14 |
| 9 | 12 | 13 | 15 | 8 | 13 |
| 15 | 11 | 12 | 16 | 14 | 16 |
| 13 | 14 | 13 | 16 | 11 | 13 |
| 23 | 16 | 14 | 15 | 12 | 15 |
| 16 | 14 | 8 | 12 | 12 | 14 |
| 12 | 14 | 20 | 18 | 11 | 16 |
| 20 | 18 | 15 | 20 | 19 | 17 |
| 15 | 16 | 10 | 17 | 13 | 16 |
| 23 | 24 | 13 | 19 | 16 | 15 |
| 14 | 16 | 19 | 23 | 10 | 12 |
| 21 | 19 | 10 | 16 | 11 | 12 |
| 19 | 18 | 13 | 19 | 14 | 15 |
| 16 | 14 | 17 | 19 | 17 | 15 |
| 9 | 12 | 16 | 14 | 10 | 10 |
| 18 | 23 | 16 | 20 | 15 | 15 |
| 8 | 13 | 9 | 13 | 7 | 12 |
| 10 | 12 | 8 | 10 | 12 | 15 |
| 10 | 13 | 10 | 12 | | |

## TABLE 51
Results of duplicate analyses on capillary blood from pre-school children
(lead concentrations in $\mu$g/100 ml)

| Scientific Officer | Institute of Child Health | Scientific Officer | Institute of Child Health |
|---|---|---|---|
| 23 | 27 | 28 | 23 |
| 27 | 19 | 21 | 23 |
| 32 | 32 | 13 | 19 |
| 23 | 23 | 13 | 17 |
| 18 | 17 | 16 | 19 |
| 10 | 17 | 21 | 23 |
| 15 | 19 | 16 | 19 |
| 17 | 19 | 13 | 15 |
| 12 | 19 | 15 | 19 |
| 17 | 19 | 22 | 29 |
| 12 | 15 | 15 | 11 |
| 13 | 11 | 20 | 21 |
| 18 | 15 | 33 | 27 |
| 22 | 19 | 24 | 17 |
| 19 | 25 | 17 | 15 |
| 20 | 17 | 12 | 15 |
| 15 | 13 | 25 | 27 |
| 34 | 21 | 27 | 21 |
| | | | |
| 17 | 13 | 30 | 21 |
| 21 | 19 | 36 | 13 |
| 9 | 15 | 18 | 19 |
| 15 | 19 | 8 | 15 |
| 12 | 19 | 26 | 25 |
| 18 | 19 | 18 | 13 |
| 17 | 13 | 23 | 21 |
| 19 | 19 | 21 | 21 |
| 12 | 15 | 17 | 23 |
| 24 | 21 | 17 | 21 |
| 20 | 19 | 11 | 17 |
| 29 | 32 | 24 | 19 |
| 23 | 29 | 25 | 25 |
| 19 | 15 | 16 | 15 |
| 16 | 17 | 15 | 13 |
| 20 | 21 | 10 | 13 |
| 49 | 50 | 12 | 11 |
| 16 | 13 | 18 | 17 |
| 23 | 21 | 19 | 21 |
| 23 | 27 | 14 | 17 |
| 20 | 15 | 16 | 19 |

## TABLE 52
Statistical comparison of blood lead results ($\mu$g/100 ml) obtained by
Scientific Officer (SO) and Institute of Child Health (ICH)

| | ICH | SO |
|---|---|---|
| *School children (venous samples)* | | |
| Mean | 14·6 | 16·4 |
| Standard deviation | 4·5 | 4·1 |
| Number of samples | 275 | 275 |

Mean difference = 1·8.
$t$-paired with 274 degrees of freedom = 9·98, P < 0·001.

| | ICH | SO |
|---|---|---|
| *Pre-school children (capillary samples)* | | |
| Mean | 19·5 | 19·4 |
| Standard deviation | 6·9 | 5·9 |
| Number of samples | 78 | 78 |

Mean difference = 0·1.
$t$-paired with 76 degrees of freedom = 0·16, P = not significant.

## TABLE 53

Comparison of venous and capillary blood lead concentrations ($\mu$g/100 ml)

| Venous | Capillary | Sex |
|--------|-----------|-----|
| 16 | 24 | F |
| 9 | 15 | F |
| 18 | 18 | F |
| 14 | 23 | F |
| 32 | 33 | M |
| 22 | 34 | F |
| 30 | 45 | M |
| 13 | 16 | F |
| 20 | 36 | M |
| 10 | 13 | F |
| 23 | 36 | M |
| 34 | 33 | M |
| 12 | 22 | F |
| 25 | 27 | M |
| 15 | 27 | M |
| 15 | 23 | F |
| 34 | 29 | M |
| 19 | 19 | F |
| 20 | 20 | F |
| 24 | 21 | F |
| 15 | 10 | F |
| 15 | 10 | F |
| 15 | 19 | M |
| 39 | 37 | M |
| 15 | 15 | F |
| 26 | 30 | M |
| 28 | 22 | F |
| 25 | 19 | F |
| 15 | 17 | F |
| 12 | 13 | F |
| 37 | 28 | M |
| 28 | 30 | M |
| 27 | 33 | M |
| 24 | 30 | M |
| 14 | 15 | M |
| 15 | 16 | F |
| 19 | 17 | F |
| 33 | 32 | M |
| Mean 23·9 | 21·2 | |

$t$-paired with 37 degrees of freedom = 2·63, $p < 0.05$.

108

# REFERENCES

Archer, A. and Barratt, R. S., 1976a, *Roy. Soc. Health J.*, **96**, 173.

Archer, A. and Barratt, R. S., 1976b, *Science of the Total Environment*, **6**, 275.

Baker, E. L., Folland, D. S., Taylor, T. A., Frank, M., Peterson, W., Lovejoy, G., Cox, D., Housworth, J. and Landrigan, P. J., 1977, *New Eng. J. Med.*, **292**, 260.

Barltrop, D. and Killala, N. J. P., 1969, *Arch. Dis. Childh.*, **44**, 476.

Barltrop, D., Strehlow, C. D., Thornton, I. and Webb, J. S., 1975, *Postgrad. Med. J.*, **51**, 801.

Berlin, A., del Castilho, P. and Smeets, J., 1972, Commission of the European Communities/ United States Environmental Protection Agency International Symposium on Environmental Health Aspects of Lead, Amsterdam, 2–6 October, paper 92.

Branquinho, C. L. and Robinson, V. J., 1976, *Environmental Pollution*, **10**, 287.

Butler, J. D. and MacMurdo, S. D., 1974, *Intern. J. Environmental Studies*, **6**, 181.

Butler, J. D., MacMurdo, S. D. and Middleton, D. R., 1974, in Transactions of the Association of Public Health Inspectors Environmental Health Congress, Torbay, p. 69.

Cernik, A. A., 1974, *Brit. J. Industr. Med.*, **31**, 239.

Cerquiglini-Monteriolo, S. and Funiciello, R., 1972, Commission of the European Communities/ United States Environmental Protection Agency International Symposium on Environmental Health Aspects of Lead, Amsterdam, 2–6 October, paper 86.

Chisholm, J. J. and Harrison, H. E., 1956, *Pediatrics*, **18**, 943.

Colacino, M. and Lavagnini, A., 1974, *Water, Air and Soil Pollution*, **3**, 209.

David, O., Clark, J. and Voeller, K., 1972, *Lancet*, **ii**, 900.

David, O., Hoffman, S., McGann, B. and Sverd, J., 1976, *Lancet*, **ii**, 1376.

Day, J. P., Hart, M. and Robinson, M. S., 1975, *Nature (London)*, **253**, 343.

Day, A. G., Evans, G. and Robson, L. E., 1977, Environmental Impact Study of an Urban Motorway, City of Bristol Environmental Health Department.

Department of the Environment, 1974, Lead in the Environment and its Significance to Man, Pollution Paper No. 2, HMSO, London.

Duggan, M. J. and Williams, S., 1977, *Science of the Total Environment*, **7**, 91.

Einbrodt, H. J., Rosmanith, J., Dreyhaupt, F. J. and Schroder, A., 1975, *Zbl. Bakt. Hyg. B.*, **161**, 38.

Farmer, J. E. and Lyon, J. D. B., 1977, *Science of the Total Environment*, **8**, 89.

Fisher, A. and LeRoy, P., 1975, *Clean Air*, **9**, 56.

109

Goldwater, L. J. and Hoover, A. W., 1967, *Arch. Environ. Health*, **15,** 60.

Hartl, W. and Resch, W., 1972, Commission of the European Communities/United States Environmental Protection Agency International Symposium on Environmental Health Aspects of Lead, Amsterdam, 2–6 October, paper 84.

*Japan Environment Summary*, 1976, **4,** 4.

La Documentation Francais, 1974, La Pollution par le Plomb et ses Dérivés, Paris.

Mackie, A., Stephens, R., Townshend, A. and Waldron, H. A., 1977, *Arch. Environ. Health*, **32,** 178.

Moore, M. R., Meredith, P. A., Campbell, B. C., Goldberg, A. and Pocock, S. J., 1977, *Lancet*, **ii,** 661.

*Morbidity and Mortality Weekly Report*, 1976, **25,** 66.

National Research Council Canada, 1973, Lead in the Canadian Environment, Ottawa.

Pasquill, F., 1974, Atmospheric Diffusion (second edition), Ellis Horwood, Colchester.

Sayre, J. W., Charney, E., Vorstal, J. and Pless, I. B., 1974, *Am. J. Dis. Child.* **127,** 167.

Singh, N. P., Boyden, J. D. and Joselow, M. M., 1975, *Arch. Environ. Health*, **30,** 557.

US National Academy of Science, 1972, Airborne Lead in Perspective, Washington.

Waldron, H. A. and Stofen, D., 1974, Sub-Clinical Lead Poisoning, Academic Press, London and New York.

Wells, A. C., Venn, J. B. and Heard, M. J., 1977, in Inhaled Particles IV (ed. Walton, W. H.), Pergamon Press, Oxford and New York, p. 175.

World Health Organisation, 1977, Environmental Health Criteria, Lead, Geneva.

Printed in England for Her Majesty's Stationery Office by Oyez Press Limited
Dd 595687   K/32   4/78